THE STRUCTURE OF THE NAZI ECONOMY

THE STRUCTURE OF THE NAZI ECONOMY

BY

MAXINE Y. WOOLSTON

NEW YORK / RUSSELL & RUSSELL

HARVARD STUDIES IN
MONOPOLY AND COMPETITION

COPYRIGHT, 1941, BY THE PRESIDENT AND FELLOWS
OF HARVARD COLLEGE
REISSUED, 1968, BY RUSSELL & RUSSELL
A DIVISION OF ATHENEUM HOUSE, INC.
BY ARRANGEMENT WITH HARVARD UNIVERSITY PRESS
L. C. CATALOG CARD NO: 68-10957
PRINTED IN THE UNITED STATES OF AMERICA

To
MY MOTHER AND FATHER

PREFACE

THIS study is the product of several years' research in German statistical documents, National Socialist writings, and foreign secondary sources. A traveling fellowship from Radcliffe College enabled me to spend some time in Germany so that I might collect statistical materials and observe at close range the operation of the Nazi system. The Caroline I. Wilby Prize granted to me by Radcliffe College for my doctoral dissertation enabled me to do additional research.

A distillate of the section on Distribution of Income and Wealth has been published in the *Review of Economic Statistics*, and the section on Corporate Profits in the *Quarterly Journal of Economics*.

I should like to express my gratitude to Professors W. L. Crum and Edward S. Mason, who read the manuscript and made helpful suggestions.

M. Y. W.

CAMBRIDGE, MASSACHUSETTS
March, 1941

CONTENTS

	INTRODUCTION	3
I.	WAR ECONOMICS IN PEACE TIMES	7
II.	GOVERNMENT AND BUSINESS	25
III.	ORGANIZATION OF TRANSPORTATION	55
IV.	CORPORATIONS	63
V.	CARTELS	90
VI.	INDUSTRIAL PRICE POLICY	96
VII.	CONTROL OF FOREIGN EXCHANGE AND FOREIGN TRADE	108
VIII.	FINANCIAL POLICIES: CAPITAL AND MONEY MARKETS, BANKING AND GOVERNMENT FINANCE	125
IX.	REGIMENTATION AND CONSCRIPTION OF LABOR	161
X.	AGRICULTURAL "PLANNING"	178
XI.	NATIONAL INCOME, CONSUMPTION, AND SOCIAL WELFARE	195
	CONCLUSION: THE POWER OF THE SWORD	230
	BIBLIOGRAPHY	241
	INDEX	247

TABLES

1. Funds for the Creation of Employment 17
2. Railroad Mileage 55
3. Transportation 57
4. Private German Shipping 60
5. Profit Ratios of German Corporations According to Industrial Classes and Composite Groups . . . 72–73
6. Dividend Payments 80
7. Average Size of German Corporations 81
8. Distribution of German Imports 122
9. Decline in Import Volume 123
10. Import Volume: Decline in Selected Groups of Commodities 123
11. Stock Exchange Sales Tax Receipts 144
12. Development of the Burden of German Income Tax According to Income Classes 153
13. Tax Payments of Wage and Salaried Workers in Per Cent of Earned Income 155
14. The Debt of the Reich, State, and Municipalities . . 158
15. Sources of Funds for Government Expenditure . . . 159
16. Armament Expenditure 160
17. Concentration of Ownership in Agriculture 182
18. New Peasant Holdings According to Size, 1919–38 . . 182
19. National Income and Factors Contributing Toward Increased Income 206
20. Gainfully Occupied in Old Reich, 1933–39 207
21. Structure of the German National Income, 1929–38 . . 208
22. Taxes and Compulsory Expenditures of a Working-Class Family in 1936 210

TABLES

23. "Strength through Joy" Activities, 1934–37 211
24. Distribution of Income in Germany According to Size . 212
25. Distribution of Wage Income 216
26. Distribution of Salaries 216
27. Distribution of Wealth in Germany, 1931 and 1935 . 218
28. Foodstuffs Available for Consumption 221
29. Building Construction in Germany, 1929–38 223
30. Cases of Sickness Reported 225
31. Pensions Paid under Social Insurance Provisions . . . 226
32. Education in the Third Reich 228

CHARTS

I. Corporate Profits, Average for All Industries, 1926–38 . 74

II. Corporation Profits for Selected Industrial Groups, 1926–38 75

III. Distribution of Income in Germany 213

IV. Distribution of Wealth in Germany 217

THE STRUCTURE OF THE NAZI ECONOMY

ABBREVIATIONS

D.V.	*Der deutsche Volkswirt*
Einzel.	*Einzelschriften zur Statistik des Deutschen Reichs*
I.f.K.	Institut fuer Konjunkturforschung
I.f.K., *Vjh.*	——, *Vierteljahrshefte zur Konjunkturforschung*
I.f.K., *Weekly Report*	——, *Weekly Report*
Rab.	*Reichsarbeitsblatt*
Rgbl.	*Reichsgesetzblatt*
Stat. d. D. R.	*Statistik des Deutschen Reichs*
Stat. Jahrb.	*Statistiches Jahrbuch fuer das Deutsche Reich*
Vjh. Stat. d. D. R.	*Vierteljahrshefte zur Statistik des Deutschen Reichs*
Voelk. Beob.	*Voelkischer Beobachter*
W. und S.	*Wirtschaft und Statistik*

INTRODUCTION

The divergent cultural tendencies growing out of the material framework of modern civilization and the thrust of dynastic politics have pushed modern man to the edge of an abyss where, huddled together with others, he seeks to assess the staggering events which have taken place in the interval of peace between two wars. Stubborn preconceptions have made this task peculiarly difficult. Western thought, blinded by the norms of an old-fashioned economy, early sought refuge in the conviction that Hitler was doomed to failure because of inflation, lack of resources, or internal revolt. What amounted to little more than a belief in magic provided a sanctuary for somewhat simpler minds. To them, the economic experiences of Germany since the advent of the Nazis have been the shadowy achievement of Dr. Hjalmar Horace Greeley Schacht and successive sorcerers. Once the spell of occult forces was broken, the German economic structure would come tumbling down, Hitler's high chair would topple over, and the rest of the world would live happily ever after.

Interlaced among these various attitudes have been diverse strands of solicitous clichés, labels to indicate the significance of the basic changes wrought by the Nazis. According to some, Nazism is "socialistic" (hence undesirable) because it subjects business and economic activity to extensive government control and leaves only the shell of private ownership. To others the Nazi system symbolizes "the wave of the future" which brings economic planning, security, and a higher standard of living for the masses.

In the face of these contradictory evaluations and the now obvious fact that neither inflation, nor internal revolt, nor lack of economic resources brought the early demise of the Nazi regime, a study of the Nazi economic structure has particular urgency. Actually the secret of Hitler's "successes" as well as the clue to the significance of the changes wrought in the economic sphere lies in the total coördination of Germany's entire manpower and natural resources — of capital and labor, of producer and consumer, of men, women, and youth — for warlike enterprise. In a bellicose philosophy of life, with its emphasis on subordination, lay the corrective for social unrest and those other disorders of civilized life which were created by depression and defeat in the first World War. Not in economic planning to raise the level of income for the enrichment of the people but in economic regimentation for military victory is to be found the distinguishing characteristic of the Nazi economy.

From the first day of Hitler's seizure of power, and more and more as each year went by and Germany approached closer to actual war, the members of the community were trained to habits of subordination and "away from that growing propensity to make light of personal authority" that is such an infirmity of democracy. The words of Thorstein Veblen in the *Theory of Business Enterprise* were prophetic: ". . . the pomp and circumstance of war and armaments, the sensational appeals to patriotic pride and animosity . . . act to rehabilitate lost ideals and weakened convictions of the chauvinistic or dynastic order. At the same stroke they direct the popular interest to other nobler, institutionally less hazardous matters than the unequal distribution of wealth or of creature comforts. . . . Loyal affections gradually shift from the business interests to the warlike and dynastic interests. . . . This may easily be carried so far as to sacrifice the profits of the business man to the exigencies of the higher politics. . . . Authenticity and sacramental dignity belong neither with the machine technology, nor with modern science, nor with business traffic. In so far as the aggressive politics and the aristocratic ideals currently furthered by the business community are worked out freely, their logical outcome is an abatement of those cultural features that distinguish modern times from what went before, including a decline of business enterprise itself."

The "logical outcome" of the system of business enterprise, so vividly described by Veblen more than three decades ago, finds its concrete expression in Germany under the Nazis. To analyze the economic anatomy of this "new order" is the specific purpose of the following pages.

Throughout, an effort has been made to rely as much as possible upon official facts and statements, and the questions concerning the validity of the data must be anticipated. Because of censorship and the widespread use of government propaganda there is an understandable hesitancy on the part of many people to accept any conclusions based on German data. But business men and bureaucrats require accurate statistics upon which to base their activities. Moreover, the statistics showing what has actually happened frequently contradict the claims of the government as to what has been accomplished, and this contradiction in itself is a partial endorsement of the accuracy of the published figures.

Whether statistics are published by democratic or totalitarian governments, painstaking criticism of the nature of statistical data — their definitions, units, consistency, comparability, and manner of presentation — is necessary for trustworthy economic interpretation. After working carefully and critically with these figures, one discovers that concealment frequently takes the form of not publishing certain economic facts. A statistical series available up to 1934, for example, will no longer be published or the specification of the numerical items will be changed from year to year so that they are no longer comparable and must be rejected. Current figures which might have a direct military significance are drawn up so as to conceal a part of the truth — that is, summary numbers are used which make interpretation difficult.

In any case, it is unlikely that falsification could be carried on consistently over a period of years. Falsified figures in one area require changes in others in any one year and demand further misrepresentation over time. This danger of course makes the task of the research worker more difficult, but the impression of general consistency indicates that the statistics are fairly reliable if used carefully and critically.

I

WAR ECONOMICS IN PEACE TIMES

THE ANALYSIS in this chapter is concerned with the development of the German recovery policy from a program aimed at the revival of private initiative into a planned economy directed toward military activity. The totalitarian state even in peace times was a preparation for the totalitarian war, and economic planning in the totalitarian state was thus restricted, for it meant economic preparedness for war. In order to give the proper perspective to the Nazi public works and armament program, a short résumé of pre-Hitler measures will also be presented.

PRE-HITLER MEASURES

State invasion into the economic sphere could go so far in 1933 because the program of the politically victorious Nazis met with demands arising from the long-continued depression and the almost colonial subservience of the German economy to foreign control.[1] Agricultural markets were in a chaotic

[1] Dr. Schacht in particular insisted that payment of reparations could not be continued if it meant increasing foreign participation in German business. "The national currents and social problems of modern countries are such that large-scale alienization of German industry would produce nationalist and social reactions which would make peaceful conduct of foreign business impossible" (Hjalmar Schacht, *The End of Reparations*, New York, 1931, pp. 25–26).

A study made by W. Salewski, *Das auslaendische Kapital in der deutschen Wirtschaft* (Berlin, 1930), shows that foreigners owned at least 50 per cent of the nominal capital of more than 430 German firms. If the daughter institutions of these firms are included, the number of firms owned and controlled by foreigners increases to more than 630. Usually foreign ownership in large concerns is known, but it cannot be ascertained for

condition. Large numbers of unemployed were thrown by the economic crisis into disillusionment and fundamental disappointment. The inflation as well as the economic crisis had destroyed the economic security of the urban middle classes. Despite the fact that monopoly power was greater than in any other country, in the face of the depression the cartel system could not prevent price competition or increase its power over economic activities without outside assistance and coercion of the government. German entrepreneurs were eager to see labor unions and the system of collective bargaining abolished and replaced by a regulated labor economy modeled to suit their interests. The spirit of nationalism which characterized the Nazi movement appealed strongly to the national feelings of the rural masses and the petty bourgeois and was unusually attractive to industry and banking involved in international difficulties.

International difficulties were everywhere accentuated by the post-war reconstruction, particularly in defeated Germany. Following the stabilization of the currency in 1924, Germany still bore the burden of reparations. The rationalization movement might have enabled German goods to be sold in foreign markets at low prices and hence have made it possible for Germany to pay for reparations as well as private loans. This trend of events, however, was made impossible because protective tariffs were erected against German goods. Moreover, with the approach of the world crisis, foreign lenders withdrew capital,[2] and markets further closed against German imports.

---

smaller concerns or for partnerships and individual proprietorships. The extent of foreign control, as estimated by Salewski, is seriously underestimated for other reasons: control by proxy and management devices are not included, and participations in selling organizations and associations, even when these are in the form of corporations or joint stock companies, are excluded. Control may also be possible without 50 per cent ownership.

[2] Dr. Schacht, fearing this reflux of funds, had attempted to stem the flow of foreign funds into Germany. He received no support from banks

At the same time the value of gold increased, so that, while the markets for German goods shrank, the quantities which had to be sold to pay reparations — loans were not obtainable — grew greater and greater. By the winter of 1931 there was the additional difficulty that certain countries, which had gone off the gold standard, enjoyed an export premium.

Germany was repeatedly advised to meet the situation by manipulating the value of her currency, but many considerations weighed heavily with the government against devaluation. The memories of the inflation were still fresh in the public mind, and the psychological factor was particularly important under a weak government. It was feared that, the moment a second depreciation began, the whole German nation would part with its marks, capital leave the country, and everyone calculate in gold as in 1922–23. In any case, the Bruening government considered the obstacles to devaluation as decisive and adopted the alternative course of accelerating the deflationary process. In order to balance the budget, important measures had already been taken in 1930. A large number of new taxes were imposed, along with an increase in the rates of existing taxes. Unemployment benefits were reduced, but, even so, relief expenditure rose because of the rapid growth in unemployment.

In the face of trade union opposition, the Bruening government enacted further deflationary measures under the emer-

---

or municipalities and was hampered by banking and currency technicalities. Under the Dawes Plan the statutes of the Reichsbank were subject to foreign control, and any large expansion of credit resulting from reduced internal discount rates would have faced opposition and criticism from outside. The note issue of the Reichsbank was restricted by law to a minimum gold and foreign exchange coverage of 40 per cent. The absence of effective open-market powers limited credit creation and prevented any direct influence on the long-term rate of interest. Whenever foreign and domestic rates were reduced, funds flowed abroad and credit markets were tightened rather than eased.

gency decrees of December 8, 1931. As a result of these decrees the average level of hourly wage rates in July 1932 was 20 per cent below the peak level of 1930. All prices regulated by cartels and the prices of proprietary articles, as well as house rents, railway fares and freights, and municipal services, were cut 10 per cent. Interest rates on all types of bonds were reduced by at least 2 per cent. Price cuts were carried through by a newly appointed Commissioner for the Supervision of Prices who exerted a downward pressure on unregulated prices. The cost of living fell about 8 per cent within three months after his appointment.

The enormous increase in tax burden contributed to the substantial decline in industrial profits. The more profits declined, the stronger became the pressure to reduce risk by eliminating small units, thus concentrating on large-scale industry. Concentration, in turn, strengthened the groups who desired to increase monopoly control by government protection.

The immediate effect of the drastic deflationary policy was mainly to accentuate the depression. Industries were very unevenly affected by the general cuts in wages and prices which took no account of special circumstances. Every contraction in employment brought on further contraction of income and employment elsewhere. Increased taxes led to a new sharpening of the depression and a new decline in revenue.

The bottom of the slump in industrial and economic activity finally came in August 1932, but the deflationary policies had been modified in May by von Papen, who had succeeded Bruening. The von Papen government proceeded at once to issue tax remission certificates which operated as rebates on the payment of certain taxes and followed up this measure with a small public works program and other devices aimed at stimulating investment and increasing employment.

In the summer of 1932 von Papen allocated 740 million RM for expenditure on housing, house repairs, land improvement, and capital expenditure by the railways and the post office. This was followed under the succeeding von Schleicher government in December 1932 by the Urgency Program for the expenditure of 500 million RM on roads, housing, public utilities, and inland water transportation. Neither the nature of the public works nor the amounts involved were unusual, and only a small part of the funds had actually been used by May 1933.[3]

It is impossible to determine exactly the total effect of the public works and other schemes put into operation by the two governments preceding Hitler; the direct and positive effects, however, were very small. The chief accomplishment of the deflationary program was to clear the way for the later spending program by lowering costs. The index of production rose from its lowest point, 60.0 in July 1932 to 62.1 in December 1932 and to 68.5 in May 1933. Stock exchange share prices also rose slowly after July 1932. There was practically no change in employment. While the issuing of tax certificates improved liquidity in the business world, it merely tended to offset the deflationary effects of a process of credit contraction which was still continuing.

It is obvious that these programs failed to improve the situation fast enough to mollify the discontent which had grown up in the depression. A bold policy of public expenditure would have run the risk not only of wiping out business confidence — thus reducing further the very low marginal efficiency of capital and at the same time further increasing the liquidity preference of capitalists — but also of raising prices so far as to upset equilibrium by the adverse balance of payments. Vested interests, fearing the possibility of a revolution growing out of the depression, and at the same time reluctant to give wide scope to a democratic government lest it open the road to socialism,

[3] Kenyon Poole, *Financial Policies in Germany, 1932–1939* (Cambridge, Mass., 1939), carefully analyzes various measures taken to alleviate unemployment.

joined other groups in overthrowing the institutions of political democracy.

WORK CREATION AND SUPPOSED REMEDIES FOR UNEMPLOYMENT

The Nazi government inaugurated as swiftly as possible a host of national economic activities, with the object of reducing unemployment. On May 1, 1933, Hitler outlined his policy of a Four-Year Plan for the abolition of unemployment. This plan was superseded in September 1936 by a second Four-Year Plan. The Nazis took the view that, unemployment being a Hydra-headed monster, the attack must be as varied as the enemy itself. Accordingly, the various unemployment programs for the years 1933 and 1934 included a number of measures, among which, on the one hand, were large public undertakings in roadmaking, land improvement, the railway and postal systems, waterways, and the like; and on the other, as an essential part of the scheme, the granting of subsidies to private building operations, and rebates of taxation on the renewals of industrial equipment.[4] There were certain other measures which had the effect of removing workers from industrial production and placed the burden of relief upon the employed by "spreading of work." The next stage came at the end of 1934, with the growing emphasis placed on German rearmament. This program, unfortunately, obscured the first effects of the work-creation measures and forcibly and *deliberately* hindered some of their secondary effects.

Supposed Remedies for Unemployment — The measures aimed at reducing the supply of labor in the ordinary labor market — the spreading of work, and the absorption of workers

[4] The most important programs following the Papen program of September 4, 1932 (*Reichsgesetzblatt*, 1932, I, 524) and the "Sofort" program of December 15, 1932 (*Rgbl.*, 1932, I, 543), were the first Reinhardt program of June 1, 1933 (*Rgbl.*, 1933, I, 323), and the second Reinhardt program of September 21, 1933 (*Rgbl.*, 1933, I, 651), with numerous amendments and executive orders.

outside industrial production by means of the labor service, the "year on the land," marriage loans, tax remissions for female domestic servants, etc. — must be regarded as "supposed" remedies for unemployment, for they provided no remedy for the waste of potential real income and wealth which results from underemployment of productive resources. Although the number of individuals employed increases as a result of such remedies, there is no tendency to increase the amount of work done. Perhaps some of those who found themselves better off spent freely to make good the privations of early years, but this has to be placed against the decreased spending of those who were taken off the labor market or whose hours of work were reduced. The chief reason for the revival of employment must thus be found in the reaction against the severe deflation of the preceding years, a reaction which began to show itself in 1932.

The nature of these "supposed remedies for unemployment" has been quite generally misunderstood. Some reports have insisted that reduction of working hours and restrictions on the dismissal of workers was a considerable burden on entrepreneurs — insinuating that employment was increased at their expense. Labor costs may have been increased as a result of the reshuffling which took place, but the most direct effect was to reduce the incomes of certain workers and increase that of others. When the campaign for the widespread adoption of the forty-hour week (with forty hours' pay) was launched in the spring of 1933, no attempt was made to introduce uniform regulations for industry as a whole. The reductions in working hours were made either as a result of negotiations between industrialists and government representatives or by spontaneous decisions of employers.[5] In any case, practically all these measures had been repealed by 1935.

[5] For example, in the iron and steel industry, the employers' association of northwestern Germany advised its members to reduce the working week

There were other measures similar in duration and character; they lasted only a short while, being repealed under the duress of an armament program, and were designed to alleviate unemployment merely by distributing the distress among the working population. Although political patronage in the form of favoritism to Nazi party members played a role in all these remedies, no clear-cut case can be made for the conclusion that they worked entirely to the detriment of the business man. For example, although every employer was required to examine the make-up of his staff with the object of ascertaining whether the age of his workers or salaried employees was in accordance with the national interest (which required that workers of a certain age, and particularly those with large families, should have employment in preference to young workers and salaried employees under twenty-five) these adjustments were to be made only after due regard for the requirements of the undertaking from the technical and economic points of view.[6] Moreover, where young employees were replaced by those over forty who had been unemployed for long periods, the employment

---

to forty hours; this was done by the affiliated regional organizations for the districts of Rhenish Westphalia, Nordmark, the Rhine, Greater Duisberg, the Remscheid area, Anhalt, Bochum, Dortmund, and Essen (*Frankfurter Zeitung*, August 22, 25, 29, 1933).

Even before the iron and steel industry had decided on this step, the employers in the lignite industry of central Germany had undertaken to introduce the forty-hour week in their plants. This again was a weekly average with provision for overtime and possible Sunday work. Adjustments were made in similar fashion in the following industries: salt, automobile, brewery, nitrates, textiles, printing, chemicals, and mining.

[6] This decree obviously increased government intervention and was very unpleasant to numerous business men. Earlier attempts of the employers' associations to encourage the replacement of younger men by the older unemployed led to numerous difficulties and to a feeling of insecurity among both employers and employees. For these reasons, the Minister of Economic Affairs issued an order on August 10, 1935, providing that the president of the Employment Exchange should have full and sole power to regulate this redistribution of employment (*Reichsarbeitsblatt*, I, August 25, 1934).

WAR ECONOMICS IN PEACE TIMES

exchanges granted in each case a monthly subsidy for six months as compensation for lowered efficiency. In March 1935 a new decree relaxed the earlier order, and young men could be allotted to vacancies, provided they had already served their term in the Labor Service.[7]

It is also unlikely that the prohibition of "multiple" earnings increased business costs to any significant amount. The main effect of this restriction was to redistribute the family income of the working population. In any case, this prohibition had been removed by 1935. Generally, multiple earnings covered not merely multiple employment in the sense of a number of different occupations carried on by the same person, but also cases where several members of the same family were together "earning more than was necessary for its own requirements." Employers were encouraged to discharge married persons whose husbands or wives were also earning wages, and likewise to dismiss unmarried women whose parents were able to support them. Multiple earnings were also presumed if, in addition to the wage, the worker or salaried employee had any other income, whether derived from an annuity or pension, a subsidiary occupation, or an independent activity, or if he possessed property.[8]

Other "supposed" remedies of unemployment which do not require special emphasis were as follows:

a. Draining of the youthful unemployed to spheres of activity with a social rather than a commercial basis, such as the Labor Service Corps, the Land Help, and the Land Year.

b. Reintroduction of universal obligation to military service.

c. Payment of marriage bonuses up to 1,000 RM to each newly married couple, on condition that the wife did not resume employment.

d. Granting of tax remissions to employers of domestic servants, and other efforts to take women off the labor market by shoving them into domestic work.

[7] *Rab.*, April 15, 1935.   [8] *Frankfurter Zeitung*, August 14, 1933.

The measures which have been discussed played only a secondary role in the recovery program. The features which were of primary importance were public works expenditures, subsidies, and tax alleviation schemes.

"Tax Alleviation" — The tax alleviation schemes were clearly devoted to encouraging investment directly by private business. Industrial concerns were given tax abatements in respect of expenditure on repairs to their buildings. The Minister of Economics was empowered to exempt from taxation enterprises engaged in developing new processes of national importance. Sums expended out of profits for replacements and renewals in industry and agriculture were exempted from income tax, corporation tax, and the tax on trading profits.[9] According to the Finance Minister the total loss of revenue from tax reductions and exemptions during 1933 was a little over one billion RM as compared to total tax yields in 1933 of almost seven billion RM. In general, however, tax rates were maintained at the very high level which they had reached during the depression years, thus only a half-hearted attempt was made to stimulate economic activity by lowering direct or indirect taxation as a whole.

Public Works — The system of public works bulked largest in the reëmployment program. This part of the program was financed by short-term government notes known as employment-creation bills, which were issued for such purposes as housing, roads, agricultural and suburban settlement, river regulation, and public utilities. Public works for the most part were not carried on by the government directly, but orders were given to ordinary firms engaged in construction work. The type of work to be undertaken, the standards set by the

[9] This provision was extended to October 1934 by allowing all expenditure on capital equipment, the estimated length of life not exceeding five years, to be deducted from the net revenue of the firm as assessed for taxation. The exemption was withdrawn at the end of 1937.

government as to execution, and the conditions under which a government order might be made were laid down in various laws and decrees. In every case the contractor had to fulfill four specifications: The project must provide for work which would not have been carried out within a reasonable time by the organizations concerned, give a large amount of employment, have usefulness for the public economy, and be carried out within a fixed time limit.

TABLE 1
FUNDS FOR THE CREATION OF EMPLOYMENT
(*To December 31, 1934*)

	MILLION RM Allocated	Expended
Reich employment-creation programs	1888	1536.5
Budget of the Reich	1135	727.6
Employment Board — relief work (Labor Service)	575	431.8
Railways	991	991
Post Office	111	111
Reich motor roads	350	166
Total	5050	3963.9

SOURCE: Karl Schiller, *Arbeitsbeschaffung und Finanzordnung in Deutschland* (Berlin, 1936), p. 155.

Some indication of the order of importance of the various public works and other programs which contributed to bring about the revival of employment can be obtained from the figures in Table 1.

The result of the combined pre-Hitler and Hitler measures was to increase "regular" employment from 12,730,000 men in June 1932 to 14,540,000 men in December 1934, and to increase substitute employment from 180,000 to 610,000 for the same period. (Regular employment covers all those engaged in ordinary employment at standard wages. Substitute employment covers those engaged in Labor Service, Land Service, and on relief works and obtaining maintenance only, hence receiving no money wages.) Total gross investment increased from 4.2

billion RM in 1932 to 8.2 billion RM during 1934. Whereas net investment in 1932 and 1933 was a negative quantity, −1.6 billion RM and −0.75, by 1934 it amounted to 2.4 billion, total investment thus exceeding the level necessary for normal replacements to be carried out. The national income rose from 45.2 billions RM in 1932 to 46.5 billions in 1933 and to 52.5 billions in 1934. An increase in income of 6 billion RM during 1933 and 1934 was thus associated with an increase of net investment of 3.1 billions.[10]

Additional figures on the secondary effects of the public works program are not available. Moreover, the initiation of an armament program obscured the effects of the work-creation measures and forcibly and deliberately hindered the expansive forces. It is thus impossible to use the German case to verify theoretical estimates that have been made of the secondary effects of public works. Judging from the contemporary improvement in economic conditions in the world as a whole, and from the fact that German business conditions were already improving at the end of 1932,[11] an appreciable measure of recovery in Germany was to be expected without the assistance of government spending. But it is very improbable that anything like the actual upswing would have taken place, having regard to the conditions prevailing in the money and capital markets in 1933.

REARMAMENT

It was early in 1935 that the work-creation program yielded pride of place as the mainspring of the drive against unemployment to the industrial boom arising from the national rearmament program. Rearmament prevented the first restorative measures from having the effect intended, according to authoritative interpretation in 1933, which was to convert a process of recovery stimulated by the state into one carried on by

[10] *Statistisches Jahrbuch fuer das Deutsche Reich*, 1938, pp. 559 and 565.
[11] Carl T. Schmidt, *German Business Cycles, 1924–33* (New York, 1934).

private initiative. It also enlarged the difficulties of foreign trade by giving rise to special import requirements and increased the anomalies which the work-creation measures had already brought about in the relative volume of production of certain commodity groups. In the course of 1935 shortages of skilled labor developed in a number of industries, especially in constructional engineering and the motor vehicle industry, and spread into other industries as the production for military purposes increased in pace.

The failure of the German public works program to revive investment outside the industrial area providing war equipment does not indict all state loan-expenditure policies. Continued and increasing government expenditure was the consequence of military and strategic considerations rather than the result of an unsuccessful attempt to bring about normal prosperity. Some National Socialists may have thought of the work-creation program as facilitating the recovery of all sectors of private business. They must soon have abandoned the attempt to carry their view into effect, for the policy actually followed was increasingly calculated to prevent an increase of private demand. Germany took drastic measures to curtail imports, particularly from the time of Schacht's New Plan in 1934, thus preventing the expansion of the consumers' goods industries.

It is sometimes contended that Germany's policy of barren investment in armaments was forced upon her because this form of investment needs comparatively little foreign raw material, while an increase in consumption would require an expansion in imports. This argument ignores the fact that part of the products needed for rearmament might have been exported and thus have provided a source of foreign exchange with which to purchase foreign consumers' goods. The scarcity of foreign exchange in Germany, with her high level of unproductive investment, was the direct effect and not the cause of the armament program.[12]

[12] Slackening of German armament production might have had bad

The "employment" multiplier, the ratio of the increment of total employment associated with a given increment of primary employment in production goods industries, was unusually low in the German case because the total effect of Nazi policies was to decrease the ratio of wages to profits to a greater extent than in the pre-Hitler prosperity. The result of this was probably to increase the addition to savings out of an addition to income.[13] The fact that the inequality of the distribution of income according to size class increased also points in the direction of a diminished propensity to consume.[14] Consumption was further limited by taxation, dividend limitation, and restriction of imports. It is here that we recognize the specific totalitarian character of the German recovery, for the policy of abnormally low wages could not have been carried through in a modern democratic country.

The government, eager to get immediate results, designed the work schemes so as to obtain the maximum amount of employment for a given expenditure of public money. As much weight as possible was placed on immediate direct employment, thus reducing indirect primary employment.

Private activity, however, expanded as far as the state would permit.[15] This is all the more remarkable when compared with the French experience during 1933–37. In France, state loan expenditure amounted to about the same proportion of national income, interference with entrepreneurs was believed to be much less, and yet private activity failed to respond to the same extent that it had in Germany. Some tentative suggestions may be offered in explanation:

---

effects in other countries, in spite of the fact that her foreign trade had been shrinking as a result of efforts toward self-sufficiency. This was because the central control of foreign trade offered a powerful weapon for dumping.

[13] See section on Corporate Profits and Dividends, Chapter IV.
[14] See section on Distribution of Income, Chapter XI.
[15] Thomas Balogh, "The National Economy of Germany," *Economic Journal*, September 1938, pp. 461–497.

1. Disinvestment by capital exports was impossible because of the strict control of foreign exchange transactions. These controls may thus be looked upon as a *sine qua non* of success.
2. Risks were considerably eliminated by regulation of prices and costs. Entrepreneurs received substantial incomes without performing the function of risk-bearing. This discrepancy between performance and reward may threaten the existence of entrepreneurs by arousing resentment in the left wing of the Nazi party. But the maintenance of titles to, if not unrestricted enjoyment of, property creates a favorable attitude toward the regime.
3. Fear of the administrative penal system probably prevented sabotage by business men.
4. Profitability of private investments undertaken to assist in the self-sufficiency campaign of the Second Four-Year Plan was ensured by the government. The government, however, dictated the rate of such investments.

The Second Four-Year Plan — By the autumn of 1936 the success of the First Four-Year Plan — the work-creation program introduced in 1933 — was no longer in doubt. Not only had unemployment ceased to be a serious problem, but bottlenecks had even appeared in building and engineering. Moreover, the high level of output increased the need for imported raw materials — the need which ever since 1934 had been the prime mover in extending state control and interference in the working of the economic system. At this juncture the Second Four-Year Plan was set up under the vigorous leadership of General Goering. The plan proposed not only to lavish large sums on the army, navy, and air force, but also to make Germany independent of all foodstuffs and raw materials necessary for military victory. It was organized under six major categories, each with an executive staff made up of army officers and big business men. They were to be concerned with:

1. Increasing the output of raw materials.
2. Distributing all raw materials, especially iron and steel, so that armament and other key industries would be able to fill their requirements.

3. Distributing labor to favor the needs of military and economic armament industries.
4. Increasing agricultural production still further, especially production of raw materials used in industry.
5. Keeping prices and wages stable.
6. Controlling and distributing foreign exchange.

Self-sufficiency was fostered in a number of ways. The government induced industries to set up or participate in the erection of many plants for the production of textile fibers, plastic and synthetic rubber, for the extraction of oil from coal and lignite, and for the increased utilization of domestic ores, such as magnesium and aluminum. Materials of which supplies were short (notably iron and steel) were replaced by cement and other materials which were relatively plentiful, and new technical devices enabled these substitutions to be made. House-to-house collections brought in empty tins, used toothpaste tubes, and other metal wastage. Snakeskin shoes were replaced by fishskin shoes; powdered eggs were made from fish protein; and Berlin busses run on illuminating gas instead of scarce gasoline. Glass was used for pipes and insulation in place of the metals Germany lacked. The reworking of old rubber, purification of used oil, and treatment of metal surfaces against rust were also a part of the intensive campaign on waste. Wastepaper was saved and reworked to conserve the paper supply; rags and waste materials — for example, bones, garbage, human hair, razor blades, horse chestnuts, and old coffee grounds — were carefully saved.

In addition, the government put on an intensive drive to maintain agricultural production and cover the two major German deficiencies: (1) fodder and feed, on which the production of livestock and dairy products depend, and (2) edible fats, including vegetable oils, lard, and butter. Summer forage was cut and stored in fermenting bins, thereby saving on expensive winter dry fodder; sawdust was stored up for manufacture

WAR ECONOMICS IN PEACE TIMES 23

into a wood flour used as fodder; bread was made partly of cellulose and recommended for obesity; sausage casings were produced from cellophane; potatoes were converted into starch, sugar, and medicinal syrup, and fats produced from flax and grape-seed.

A paradoxical comfort is to be found for armaments in the fact that the accumulation of this type of investment goods does not decrease the rate of profit; the tragedy of useful investments is that they are created to serve the needs of man by being consumed and hence stale with abundance. Moreover, the production of military weapons, instead of competing with private interests, enables profits to increase as a result of additional purchasing power poured into the market by the government. Armament programs thus have the special advantage of yielding employment opportunities in the producers' goods industry without being endangered by the failure of a revival in consumer demand.

In the long run the outlook is less rosy. Real income falls short of the amount which should have been obtained had the factors of production been used directly or indirectly to make consumers' goods. Once full employment is approached, armament production can only be increased by lowering consumption. Until that time, of course, it is possible to expand without lowering living standards. Beyond that point, restriction of consumption will occur in one or several of the following ways: inflation, increased taxation, low wage policy, rationing. Inflation becomes likely only as the limits of the other three possibilities are approached. But no danger can arise, even if the whole of public expenditure is financed by creation of cash, so long as state loan expenditure at full employment is kept within the capacity of the productive system — having due regard to the enforceable minimum of consumption and private investment.

Although an armament program increases employment and profits in the short run, from the longer point of view any system of economic planning which revolves about an armament axis must be classified as restrictive rather than expansive. The goal of capital accumulation and the continuation of profits is carried out through imperialist conquest rather than by the solution of the basic problems from which the depression grew.

II

GOVERNMENT AND BUSINESS

THE TENDENCY toward increased state intervention in post-war years has made the relationship between the state and the economy not only one of the most complex problems of modern society but also one of the most important. Two separate although intimately related questions are involved in this relationship: To what extent and by what means is the totalitarian state used by different economic groups as a weapon? To what extent and by what means does the state direct, influence, or control economic activities? Hitler himself boasts that unlike the pluralistic democratic state, the totalitarian government is so strong that the leader can reach his decisions free from the pressure of any group.

There are two ways of testing this boast: first, by an analysis of the economic policies of the state, from which inferences concerning the forces at work in the formulation of these policies may be drawn more or less accurately, and, second, by a careful examination of the institutional controls and the personalities directing them.

Economic policies have been partially dealt with in the earlier chapter on the recovery and rearmament program, and the benefits and losses of various groups in the economy will be referred to later on when the distribution of income is considered. Special chapters will be devoted to the consideration of cartels and the corporation. Institutional controls pertaining to business in general are examined here and the regulatory mechanisms in the fields of transportation, labor, agriculture, and money and banking will be analyzed in the chapters immediately following.

The investigation aims primarily to examine where power lies; descriptions of the institutional structure are given only in so far as they relate to this central question. A complete testing is not possible because the data are insufficient — it would be necessary to know the biography of every German official intimately enough so that his relations with business men, party members, military corps, and labor could be judged. Yet certain things can be said about the personalities holding the strings of control at strategic points. In addition, certain information can be gained from the activity of various groups at critical times in economic development and during the process of institutional changes.

Instead of directing the economic system from within by state ownership of certain key positions in the economy, the Nazis enforced regulation from without through cartel, price-fixing, labor and banking legislation, and administrative control devices. That this more conservative principle would be employed in the state's control of the economy was predetermined by the factors surrounding the rise of the National Socialist movement. It is true that the powerful capitalist groups had not created the movement, but they gave it financial support in a crucial period. The anti-capitalist points in Hitler's platform at first aroused considerable skepticism. But when Hitler, invited to interpret his program before the associations of industrialists, stressed "the necessity of private property and of an economic order based upon the profit system, individual initiative, and inequality of wealth and income," the conservatives, particularly the producers of iron and coal, became convinced that an opportunity for them lay in the National Socialist movement.

The industrialists of the Ruhr understood Hitler's statement that "the system corresponding to political democracy in the economic field is communism" to mean an offer to help destroy the Weimar democracy, and they welcomed him as an agent for wiping out the dangers of socialization threatened by political democracy. As a result of the economic strain of the post-war period and the world crisis, the state threatened to dominate the economy. Thus, permanent control over the government appeared as a life and death matter to the propertied classes. Two powerful groups, especially, were politically united in their desire for an autocratic regime over which they had control: the steel and coal industrialists on the Rhine and the Ruhr, and the land-owning aristocracy of the east. Leading men among these two groups were convinced that the social democracy of the Weimar Republic was in the long run irreconcilable with their own social existence.[1]

To establish a dictatorship without or against Hitler, whose followers numbered more than ten million people and several hundred thousand armed men, seemed a difficult task. The Schleicher and Papen attempts had failed. Although very little accurate information is available regarding the downfall of these presidential cabinets, it is clear that the majority in conservative circles preferred a solution under Papen to a solution under Hitler. It was only after several attempts to replace the democratic state by an autocratic government had failed, that Hitler, the leader of a powerful mass movement, could figure as a serious alternative to Papen, Schleicher, and other would-be dictators.

TRENDS IN GOVERNMENT OWNERSHIP

In return for business assistance, the Nazis hastened to give evidence of their good will by restoring to private capitalism a number of monopolies held or controlled by the state. It is frequently maintained that this return of capital to private enterprise was of no practical consequence because meanwhile

[1] Alfred Hugenberg, *Streiflichter* (Berlin, 1927); Gaston Raphael, *Le Roi de la Ruhr Hugo Stinnes* (Paris, 1924); Gustav Stresemann, *Vermaechtnis* (Berlin, 1932).

the state had assumed full control of the economic system as a whole.

This statement is generally made without any details as to the dates on which the government transferred ownership to private hands. As we shall see later, this transference occurred early and had for the most part been completed by 1936 and before the centralization of controls brought forth as a result of the rearmament program. Moreover it had practical significance even after the extension of government intervention. For example, the return of the German Steel Trust to private ownership enabled owners to reap the rewards of rearmament. Although dividends were limited to 6 per cent and corporation taxes were increased after 1935, these factors did not prevent entrepreneurs from making profits. It is true, however, that the exorbitant profits of the first World War were not permitted.

The gigantic war program was possible only if consumption remained at a comparatively low level, or, in other words, if savings increased, and the propensity to save was facilitated by inequalities of income. Inequality of income, the partial source of which lies in inequalities of wealth, was thus secured by "reprivatization." It would have been possible theoretically for the government to have taken over private enterprise and then to have distributed returns from state enterprises unevenly among the population, but by distributing income according to ownership the Nazis received the sanction of tradition and the support of propertied interests.

The practical significance of the transference of government enterprises into private hands was thus that the capitalist class continued to serve as a vessel for the accumulation of income. Profit-making and the return of property to private hands, moreover, have assisted the consolidation of Nazi party power.

The trend of events during the Nazi years seems all the more significant when placed against the background of earlier years. State ownership and control was an old story in Ger-

GOVERNMENT AND BUSINESS

many; she entered the liberal era with a relatively large sector of state-controlled economy. On the eve of the World War the following sectors of the economy were completely government-owned: postal system, telephone, domestic telegraph, and railroads. The gasworks, waterworks, and traction system were under municipal or mixed ownership — mixed ownership companies were usually initiated by private entrepreneurs who turned over stock interests to municipalities in return for the grant of a franchise. Municipal, state, or mixed ownership forms controlled the electric power industry. States and municipalities owned a few ironworks and mines and were dominant in forestry. Even among the large-scale farms, state-owned units had some importance.[2]

Government ownership, moreover, was further expanded during the World War and the Weimar Republic. Industrial enterprises, such as nitrate and aluminum plants, the army shops, and power plants, were transferred to individual stock companies and the stocks concentrated in a Reich-owned holding company, the Viag (Vereinigte Industrie Aktiengesellschaft), with which was combined the Reichs-Kredit-Gesellschaft as the financial organization. In March 1919 the government proclaimed a socialization act as an installment payment on its earlier promises to the Socialist workers' organizations. At the same time the Law for the Organization of Coal Mining was proclaimed, followed in April 1919 by a Law for the Organization of Potash Mining.[3] In both these industries, as well as in the steel industry, government commissions were set up as a first step toward socialization.

The government of the Weimar Republic also increased its activities in the banking field. The Reichs-Kredit-Gesellschaft expanded to the rank of one of the "Big Four." Both the na-

[2] Gustav Stolper, *German Economy, 1870–1940* (New York, 1940), pp. 57–76.
[3] Stolper, *op. cit.*, pp. 200–201.

tional and state governments founded banks for special purposes — in the field of intermediary building credits, the Deutsche Bau- und Bodenbank, as well as several mortgage banks; in short-term railroad financing, the Reichsverkehrsgesellschaft; and in long-term agricultural credits, the Deutsche Rentenbank and the Rentenbank-Kreditanstalt. Finally, the Reichsbank in order to obtain greater freedom of action founded a subsidiary, the Gold Discount Bank (Golddiskontbank).

Gustav Stolper estimates that on the eve of the 1929 slump public banks (apart from the Reichsbank and savings banks) accounted for at least 40 per cent of the total assets of all banks. But, although the state held a predominant position in German banking prior to the banking crisis of 1931, the chief function of the German credit system, the financing of industrial and banking companies, until the depression remained in the hands of private banks. With the banking crisis, the Reich took over the control of capital in most of the large banks, and, as we shall see, it was not returned to private hands until the accession of the Nazis.

The trend of government ownership in various industries may now be considered. The United Steel Trust is an outstanding example of "reprivatization." In 1932 the state had obtained control of the Trust by purchase of stock, amounting to 125,000,000 RM, in the powerful Gelsenkirchen Mining Company.[4] The expenditure had been defended by democratic Reich Minister of Finance Dietrich on the grounds that this immense trust should be withdrawn from the influence of private hands for political and economic reasons. When in November 1933 a merger was proposed, the Gelsenkirchen Mining Company to take over all the assets of the Amalgamated Steel Works, of the Phoenix Mining Company, and of the Van der Zypen Steel Company — changing the name to

[4] *Economist*, July 8, 1933, p. 73.

Vereinigte Stahlwerke, A.G. — the new transaction again opened for debate the whole problem of state control over the Steel Trust.[5]

One plan proposed the complete break-up of the Trust by dividing it into small independent companies. The state, however, sold back the control of the Gelsenkirchen for 100 million marks, and the gains from the amalgamation were more than sufficient to cover interest payments and the greater part of capital depreciation. All shares owned by the various concerns were withdrawn and those in circulation exchanged for those of the new holding company. Herr Thyssen at the time of the merger promised the shareholders in the general meeting of the old steel concern that there would be no further reorganization at their expense.[6]

The policy of restoring to private capitalism the properties held or controlled by the state is also evident in the case of the banks. As a result of the bank crash of 1931, most of the big banks had come under state control. Dr. Schacht estimated before the banking inquiry committee in 1934 that the Reich controlled directly or indirectly 70 per cent of all German corporation banks.[7] The government, either directly or through the Gold Discount Bank, owned 91 per cent of the capital stock of the Dresdner-Bank and Danat (merged); 70 per cent of the Commerz- und Privatbank; 35 per cent of the Deutsche Bank und Diskonto Gesellschaft; 70 per cent of the Allgemeine Deutsche Kreditanstalt; and 66.6 per cent of the Norddeutsche Kreditbank.[8]

Notwithstanding the widespread support of socialization of the private banks by the radical wing of the party, the Hitler government pronounced itself against the nationalization of

[5] *Economist*, November 4, 1933, p. 846.
[6] *Ibid.*
[7] *Zeitschrift fuer oeffentliche Wirtschaft*, December 1934, p. 506.
[8] *Untersuchung des Bankwesens* (Berlin, 1933–34), pt. I, vol. I, p. 396.

the banking system in the report of the investigating committee in December 1934.[9] The Deutsche Bank, which as early as 1933 had got back nearly 20 million RM of its stock from the government in return for giving the state a big building, repurchased 35 million RM of shares in 1936 and completed the repurchase in March 1937, becoming a wholly private institution.[10] In October 1936 the government liquidated its holdings of the Commerz- und Privatbank to a consortium of private banks. In September 1937 an announcement was made at a directors' meeting of the Dresdner Bank that its shares were no longer held by the government.[11]

In addition, the government returned a block of shares of the Deutscher Schiff und Maschinenbau to a group of Bremen merchants in March 1936, and in September 1936 it sold 8.2 million RM of stocks (out of a total capitalization of 10 millions) in the Hamburg Sud-Amerika to a Humburg syndicate of merchants.[12]

Likewise, National Socialism opposed municipally owned enterprises which had been fairly profitable even during the depression. Until January 1, 1935, their profits had been exempt from taxation, but after that date they were required to pay the same taxes as all other corporate enterprises. The law of 1919 permitting "socialization" of power production was repealed on December 13, 1935. The expressed purpose of this legislation was to remove the "disorder" created in the distribution of electrical power by "municipal socialism," and the

---

[9] The radical wing of the National Socialist party, particularly Dr. Feder, the former Undersecretary of Economics, advocated the solution of all banking and credit problems in the nationalization of the entire German banking system (Gottfried Feder, *Das Manifest zur Brechung der Zinsknechtschaft*, Munich, 1932).

[10] League of Nations, *Money and Banking*, 1938, II (Geneva, 1939), p. 92.

[11] *Ibid.*

[12] Reichs-Kredit-Gesellschaft, *Germany's Economic Situation at the Turn of 1936/37*, p. 55.

preamble of the law explained that "such an organization of electric power production is contrary to the basic idea of the National Socialist concept. . . ." Henceforth not only were private plants to be freed of all "unnecessary impediments," but also every possible encouragement was to be given them.[13]

The effect of this policy is clearly indicated in the decline of revenue from municipally owned enterprises. Revenue from such sources had shown a constant trend upward from 1925 to 1931. This upward trend was a concomitant of two factors: the growth of ownership and variations in profits during the course of the business cycle. Although income decreased in 1931 and 1932 because of the depression, the greatest drop is seen in 1934–35 and 1935–36 — two years in which recovery was already noticeable. The decrease for those years thus reflects the attempt of the Nazis to restrict the area of municipal ownership as well as to tax the profit from such activities.[14]

Since 1936 there has been a shift from municipal and state ownership to the Reich and an increase in the capital of state and local banks and in mixed ownership companies. On March 31, 1938, the Reich, municipalities, and states owned 331 corporations with a capital stock of a little over two billion RM, representing approximately 10 per cent of the total capital stock of all corporations in Germany.[15] These figures, of course, do not include railroads or post offices. Despite the

[13] *Le Temps*, January 21, 1936.
[14] Income from enterprises owned by municipalities was as follows (in million RM):

March 31, 1926	324	March 21, 1932	515
1927	390	1933	481
1928	425	1934	494
1929	480	1935	456
1930	482	1936	360
1931	636		

Einzelschriften zur Statistik des Deutschen Reichs, 1938, Nr. 37, p. 38.
[15] *Vierteljahrshefte zur Statistik des Deutschen Reichs*, 1939, III, p. 64. Mixed ownership companies are included in this tabulation.

increase of Reich holdings in mixed ownership companies as a result of the war drive, the capital involved is still not large. Moreover, government undertakings of this type are carried on frequently to provide business with effective guarantees against losses.

The Hermann Goering Reichswerke fuer Erzbergbau and Eisenhuetten, which will be discussed more fully in a later section, may be indicative of a new trend. In this case, government financing was continued and expanded not only because of the risks involved in setting up an organization for the exploitation of low-grade iron ore but also because it gave Nazi party members an effective weapon against existing competitors in the iron and steel industry. Early in 1938, however, when the capital stock was considerably increased, the government sold a large share to private business. The Reich invested 270 million RM in the total capital stock of 400 million RM. The remaining shares in the form of preferred stock were taken over by private individuals who subscribed for them in the open market.[16] Majority control, however, is still retained by the government, which owns all the common stock. Goering, who had large holdings as a private individual, could thus use his position as a top party official to increase his own private wealth in competition with existing business interests.

At the time this company was formed, the *Bergwerkszeitung*, the newspaper of heavy industry, expressed surprise that the creation of the company was considered abroad as a measure for nationalization. "The state," it said, "spares private industry the risk of investing new capital and leaves it the responsibility of sharing voluntarily in the execution of great new projects."[17] General Hanneken, a department head in the Ministry of Economic Affairs, told the occupational group for

---

[16] C. W. Guillebaud, *The Economic Recovery of Germany* (London, 1939), note, p. 104.

[17] *Deutsche Bergwerkszeitung*, July 27, 1937.

the iron industry that "as soon as possible the Goering-Werke will be returned to private ownership." [18]

An idea of the close interpenetration of the state and private industry achieved in the "mixed" enterprises can be obtained from the supervisory board of the big Rheinmetall-Borsig (subsidiary of the Goering-Werke), composed of the following: four representatives of big capital, Herr Borsig and Karl Bosch of the I. G. Farbenindustrie; a representative from the Deutsche Bank and one from the Dresdner Bank; a representative of the old aristocracy, converted to National Socialism, the Duke of Saxony-Coburg-Gotha; two state representatives, State Secretary Trendelenburg and a representative of the Ministry of Finance; a representative of the army, General Thomas; and finally, two representatives of the Goering-Werke and one from the Reichs-Kredit-Gesellschaft.[19]

EMPLOYERS' ORGANIZATIONS

During the early period after the seizure of power, the new political masters subjected economic organizations, trade unions, and employer associations to their control. Government commissioners and many industrial enterprises, fearing further political repercussions, replaced Jewish directors and employees with Aryans related to the party movement. The fear among many businessmen that this wave of "coördination" might turn into a social revolution was removed after several months of uncertainty when the government freed itself from the radical wing of the party.

Soon after Hitler's accession to power, an impatient clamor for a corporative structure to include employers' and workers' associations was heard from the Nazi left wing. Dr. Wagener, who represented this attitude in his official capacity as head of the economic section of the National Socialist Party, got the

[18] *Frankfurter Zeitung*, June 12, 1938.
[19] *Deutsche Freiheit*, July 7, 1938.

German Federation of Industry to assume the title of Corporation of German Industry on April 6, the date of the dissolution of the former organization. Wagener demanded that President Krupp retire, but the latter remained as president of the new "corporation," assisted by two Nazi commissioners.[20] For a time, the policies of Wagener were influential, and on May 31, 1933, Hitler announced the promulgation of a law outlining the corporative structure.

A committee was appointed to draw up the statutes of the new corporative state. Every radical Nazi had his plan, and each hoped that the new structure would increase his privileges. There were Dr. Wagener's plan and the plan of Dr. Renteln, leader of the Combat League of the Middle Classes. But the plan that seemed to have the greatest chance of success was Dr. Ley's. He proposed nothing less than the absorption into his Labor Front of the entire economy, both workers' and employers' organizations: "The corporative structure of the German people," he exclaimed, "is complete in its main outlines! It is nothing less than the establishment of an organic tie between the workers and employees on the one hand, and the employers on the other, and their common integration into the economic organism." [21]

The opposition of the industrialists supported by the Reichswehr, however, resulted in the veto of this plan by July 1933, and Dr. Wagener was removed. On July 13 the new Minister of Economy, Schmitt, assured the industrialists that the corporative structure was postponed to better times because there was danger that irresponsible elements might try to make adventurous experiments in this field. Dr. Ley, who had not lost all hope of achieving his goal, continued to announce, unperturbed, the advent of the corporative state.

Hitler was forced to make a compromise between his finan-

[20] *Le Temps*, April 7, 1933.
[21] *Le Temps*, June 1, 1933.

cial backers and the forces within his party who hoped to control the employers' organizations. When the employers' economic organizations, semi-official in character, were made into occupational groups by the law of February 27, 1934, no representative of the wage earners was admitted. The radical elements were told by Minister Schmitt, the author of the law: "In the measures taken at present, there is no question yet of a corporative reform. You know that the Fuehrer has intentionally postponed the solution on this problem, for he rightly believes that a corporative structure should arise only slowly with the development of events." [22]

Dr. Ley and his following were persistent. The law of February 27, 1934, though a severe check for them, left a few trump cards in their hands. They had at least been successful in that the new employers' organizations had a pronouncedly governmental character, and the employers were in some degree regimented in them. The "leader principle" was applied, membership was compulsory, and there were no more deliberative assemblies. They hoped thus to be able somewhat to control from above the activity of the industrialists. They enjoyed two other advantages: the Federation, now Corporation, of Industry, was dismembered into several different occupational groups, and Kessler, who had been appointed Minister of Economy, seemed determined on dissolving the old private employers' associations. Once this plan was carried out, once the employers' associations were staffed and controlled by the National Socialist party and government, perhaps it would be possible to push them into the corporative structure.

But the old employers' associations, particularly the Federation of Industry, stubbornly refused to let themselves be dissolved, and on July 11, 1934, Kessler was recalled. In his turn, von Goltz, associate and successor to Kessler, tried to dissolve the recalcitrant associations, but failed and was re-

[22] *Le Temps*, March 14, 1934.

moved at the end of November 1934. On December 2, Dr. Schacht, the new Minister of Economic Affairs, assisted the employers' demands by reuniting the seven "occupational groups" of the Federation of Industry into one. Furthermore, he restored to the employers a measure of autonomy by weakening somewhat the leader principle. A general membership meeting was to be held at least once a year, and the administration of the head of the group had to be voted on by secret ballot. The outcome of this vote must be reported to the director of the next higher formation, who, in case the assembly had passed a vote of censure upon its director, must consider the latter's recall.[23] Dr. Ley, moreover, in 1934 announced that the idea of a corporate state as expressed in Point 25 of the party program had definitely been abandoned as impractical, semi-democratic, and pluralistic. This is the only instance in which the National Socialists have repudiated officially one of the points in their program.

The government commissioners for the industrial associations who had been appointed in the wave of coördination were recalled, and several of those advocating a semi-socialistic guild organization in industry and commerce were placed in concentration camps. There was no change of policy in the relation of the government and the employers to the former trade unions, and it would thus appear that, in regard to the important issue of organizational structure of the economy, business interests were at this time more influential than labor. While no labor representation is admitted in the economic organizations, councils, and commissions, there is, however, business representation in the Labor Front. At every step of the organizational structure the employers are represented twice. They belong both to their own occupational group and to the shop community of the Labor Front; on a district basis

[23] Great Britain, Department of Overseas Trade, *Economic Conditions in Germany to March, 1936* (London, 1936), p. 88.

they are then represented in the district economic commission and in the labor commission; and on a national basis they hold seats in the Economic Council of the Reich and in the Labor Council. Collaboration thus exists only in the labor organizations, commissions, and councils.

THE ESTATE OF INDUSTRY AND TRADE

By an act of February 27, 1934, the Minister of Economic Affairs was directed to prepare the groundwork for the organization of industry and trade. Under the provisions of this act the Minister linked together under his ultimate control all organizations which had at one time been formed to protect business interests. Only organizations which were recognized by him were permitted to exist, and these were all compelled to come into the Estate of Industry and Trade. Under this new set-up, business is organized along both regional and functional lines. The functional organization consists of six national branches: Industry, Handicrafts, Trade, Banking, Insurance, and Power. Within each national branch are various sub-units called trade groups, and these are by far the most important bodies in the functional organization. They collect contributions and draw up the budget for their sections. The director of a trade group is authorized to impose disciplinary punishment on members.[24]

The regional organization of the Estate of Industry and Trade has fourteen districts. In each district there is an industrial board or district group representing all economic interests, since the functional groups represent specific interests of particular lines of business and trade. These district groups are the former regional employers' associations remodeled on the leadership principle, and their directors are in practically all cases well-known industrialists. The next stage lower in the regional representation are the chambers of commerce.

[24] *Rgbl.*, 1934, I, 185; *Rgbl.*, 1935, I, 1169.

Membership in a local chamber is compulsory for all business men. The former Federation of German Chambers of Commerce was converted into the Association of Chambers of Commerce and Industry.

The functional and regional organizations in turn are combined in a National Economic Chamber (*Reichswirtschaftskammer*). All six national business branches are represented in this supreme body of the whole Estate. The legal status of the National Chamber is determined by the Minister of Economics, who appoints the director and his deputies. The director is assisted by an advisory council consisting of:

1. Director of the Combine of the Chambers of Commerce
2. Directors of the National Branches and Main Groups
3. Directors of the Industrial Boards
4. Four representatives of Transport
5. Representatives of the Reich Food Estate
6. Representatives of the Municipalities
7. Industrialists appointed by the Minister for Economic Affairs [25]

The stated purpose of the Estate of Industry and Trade was to eliminate "the excessive organization of German business hitherto prevailing, with its resulting inactivity, as well as the obstruction and disturbance caused by the rivalry of individual organizations. It is planned to carry out a comprehensive, strict and uniform organization of all parts of industry." [26] The structure which was set up did not fulfill this end in all respects; for, in spite of later alterations and simplifications, complaints against overlapping and double-work were frequent. There was a reshuffling of previously existing organizations, a simplification of former fields of control, and further centralization of direction from above.

The previously existing associations which protected business interests were fused within the new organization. But no

[25] Department of Overseas Trade, *op. cit.*, p. 92.
[26] *Rgbl.*, 1934, I, 185.

GOVERNMENT AND BUSINESS 41

rival business man has been permitted to stay out and spoil the game, for membership is compulsory and failure to register is subject to heavy fine. All employers and undertakings (natural and juridical persons, private undertakings, and those of the states and municipalities) independently engaged in trade or industry are compulsorily and automatically members of their respective functional and regional units. Central business organizations which at one time were independent of each other except for voluntary coöperation or for dual membership were coördinated by the Nazis, thus eliminating a considerable confusion of duplication and overlapping of functions.

The only other important change, the application of the director or leader principle, means further centralization along lines characterizing the growth of monopoly in the last seventy-five years. The whole structure is subordinated to the Minister of Economic Affairs, who appoints the director of the top classification in the hierarchical arrangement of control. Each division has a director who is appointed at the recommendation of the director of the next higher division, or by the Minister of Economic Affairs. Appointment, removal, and control are centralized for the whole economy in a manner similar to that found in a business corporation.

The law does not give an exact definition of the functions of the individual organization but transfers the functions and powers of the former business associations to the new units. Any attempt to make clear-cut divisions between the functions of these various formations would be attributing to them a definiteness which does not exist in the law.[27] In accordance

[27] Robert Brady, *The Spirit and Structure of German Fascism* (London, 1937), delineates their functions as follows: The National Economic Chamber is the policy-forming hierarchy; the Chambers of Industry and Commerce provide the policy-enforcing hierarchy and the third division, which he calls Trade Associations, furnish the "policy pressure hierarchies" for German business. Inasmuch as Brady does not give the German names it is difficult to transpose his designations to the corresponding groups

with the party motto, "a government which is obliged to govern too much is not worth anything," leaders for the most part get their instructions not from laws but from verbal injunctions from higher officials or from suggestions and recommendations of members. "Very often it is difficult to determine where personal volition ends and constraint begins." [28] The general formula of the act is so broad as to authorize the leaders or directors to undertake almost any action except that of price and market policies, which are the special problems of the cartels and special price-control organizations set up by the government. They give advice and assistance to their members in their special fields of trade or production, and the cases in which action is taken that amounts practically to market regulations increase steadily in number and importance.

Questions relating to general economic policy, money, banking and credit policy, foreign exchange control, taxation matters, insurance, patents, propaganda, and law belong to the Estate of Trade and Industry. Directors are entitled to carry on investigations and to issue orders as a means of enforcing the principles and economic policies of the government and to preserve the rules of honor and fairness in their branch of business. Orders of group leaders may be enforced through fines up to 1,000 marks; appeals against orders of group leaders may be addressed only to the leader of the national branch. His decision is final.

The machinery that is thus set up to supervise industry and trade can be used to achieve almost any purpose so far as the formal provisions are concerned. There are no longer any

---

used here. The Department of Overseas Trade (*op. cit.*) describes the functions only in a very general way. Henri Laufenburger and Pierre Pflimlin, *La Nouvelle Structure économique du Reich* (Paris, 1938), insist that it is impossible to determine the exact activity of the various organizations in the Estate of Industry and Trade.

[28] Laufenburger and Pflimlin, *op. cit.*, p. 17.

GOVERNMENT AND BUSINESS

formal obstacles, in view of the leadership principle, to controlling the business man lock, stock, and barrel should the Minister of Economic Affairs see fit to do so. By the same token there is nothing to prevent the business men from doing exactly as they please if they have any means for controlling the controllers. The facts of the case probably represent a compromise between these two extremes. In the majority of cases the presidents of the former associations transformed into the new groups were retained as "leaders." [29]

The general form of the Estate of Industry and Trade shows that the Nazis intend that the German economy shall be a controlled capitalist economy. Government sets the task for industry to perform and requires industry to carry out the task. Private property is emphatically declared to be part of the National Socialist scheme of things. On the other hand, private property is subject to the interests of the state, and this principle is embodied in one of the main slogans of the Party, "Gemeinnutz vor Eigennutz." The application of such a slogan must mean the subordination of the interests of the individual to what the ruling power in the state decides to be the interests of the community, and the principle appears to many critics to be contradictory to the emphasis on private property and enterprise. But private property and state regulation are not opposed to each other if the ruling power has interests identical with those of the owning part of the community, or if the ruling power is the owning group.

"Gemeinnutz," or the "common good," has on the one hand been interpreted to mean that the ends of the state in building an efficient war machine must be the dominating element within the economy. This policy means that private initiative in consumers' goods industries is restricted, but it also gives an overwhelming impetus to heavy industries which fill the state's orders for war goods. From the point of view of the business

[29] Laufenburger and Pflimlin, *op. cit.*, p. 51.

man, common good means that no one should be allowed to "spoil the market" by "unfair" or "dishonest" methods. If business policies are "fair" to business, the principle holds good that all are exercising their talents on behalf of the "common good" of business, but in addition business must not sabotage war aims.

This interpretation is indicated in a speech of State-Secretary Rudolf Brinkmann on October 21, 1938, before the National Congress of Banks and Insurance Companies. Brinkmann analyzed the functions of the Estate of Industry and Trade as follows:

> In creating this organization the state met both the natural inclination of the German for unification and voluntary subordination as well as the only too understandable wish of business to have associations or boards at its disposal which would enable it to deal with the state and which would offer it the possibility of gaining a hearing for its experience, its desires and its needs. . . .
>
> This organization has been able to make itself useful and in many cases indispensable to its members as well as to the state. It has offered its members — and thus drawn them closer and closer to it — that which every business man seeks; — promotion of his firm in all those fields where success can be achieved only by cooperative activity. I need only call your attention to the development of comparative business data, to the insuring of the assembling of business statistics necessary for such comparisons, to the education towards the use of new synthetic materials, to the re-training in their use and the alleviation thereby of many needs, to the promotion of business norms and the establishment of uniform standards of quality. Important from the state's viewpoint was the readiness with which the Estate of Industry and Trade tackled the difficulty of raw-materials and export production. There is much still to be done and the small business man should have more of a feeling that his problems are being taken care of and that he is being treated better than the large business man who can easily help himself.[30]

[30] Institut fuer Konjunkturforschung (German Institute for Business Research), *Weekly Report*, Supplement, November 2, 1938, pp. 2–3.

RESTRAINT OF COMPETITION

The activities of the Estate of Trade and Industry undoubtedly restrained competition, but this does not mean that the organization is thereby set up to the detriment of business. Restraint of competition may be exactly what certain business interests want to achieve.

But should an official statement be accepted as an accurate description of the functions of the Estate of Industry and Trade? You never judge a man solely by what he says of himself and in the same way you cannot judge a social movement by its own proclamations. In this particular case, however, there are extenuating circumstances. Brinkmann was an advocate of free competition, and this particular speech is remarkably outspoken. In 1939, three days after a similar speech, Dr. Brinkmann retired, according to the official statement, as a result of a "nervous breakdown with loss of memory." [31] Circumstantial evidence thus indicates that Brinkmann spoke with unusual candor. For this reason the following quotation from his speech of October 1938 has particular significance for an understanding of the relation of government to business.

> You will call attention to the facts that the freedom of disposition of the entrepreneur in the sphere of commodity purchases is chained down by the system of supervisory boards and other regulations, that the utilization of labor is subject to various restrictions, that the wage ceiling and prohibition of price increases force a price level which in a liberal economy would be impossible.
>
> And you will argue further that . . . there is occurring under the eyes of the state the very thing it wishes to prevent, namely, choking up of individual initiative by administrative activity, a burdening, perhaps even an overburdening, of the economic apparatus with dead costs; the impairment of a standard of living to be derived from a certain nominal income, due to rising taxes and monopoly prices; a still further expansion of the already great concerns, and death or dormancy among the small and medium-size independent business.

[31] Guenter Reimann, *The Vampire Economy* (New York, 1939), p. 162.

Of course, much of this is exaggerated and biased, but much cannot be denied. It is true, for example, that the state has had an increasing share in national income. . . . It is also true that there has been a large increase in the number of establishments which business uses to insure itself a monopoly position and the profits resulting therefrom. . . . I need not tell you that I consider freedom of action the correct course and will push it wherever I can. But how can I do it, if this freedom of action is considered a plague and not a benefit?
. . . The State helped to eliminate dishonest elements from competition by application of protective regulations, by constructive prohibitions and by protection of titles. It practically cut off all foreign competition. It interested itself in full employment and thus eased the pressure on the selling side. . . . But what is it but a subsidy out of the pockets of other fellow-citizens, when a group of entrepreneurs, supported by a powerful position, which they have created through restrictions of competition, forces prices on other sections of business or changes the cost-return relation to their advantage, and when the state grants this group a concession? . . .[32]

The core of Brinkmann's criticism is thus that there is too much state intervention on behalf of large concerns and monopolies, which has led to restriction of competition and the decline of small and medium-sized independent businesses. His insistence that government control was being exercised on behalf of larger firms and at the expense of the smaller unit is clearly seen in the activities of one of the national branches of the Estate of Trade and Industry — the national branch called "Trade."

This branch has helped the large department store and the chain store at the expense of the little retail firm — this in spite of the fact that the Nazis in their rise to power made virulent threats and hurled bitter words against chain and department stores. National Socialism was even thought of as the revolt of the little business man — revolt against encircling monopoly and predatory business organizations, revolt against the crushing domination of giant concerns. But plans to break

[32] I.f.K., *Weekly Report*, Supplement, November 2, 1938.

up the department store in accordance with the party program were never carried out, and subsidies were even given to several of them.

The chain and department stores were left unmolested except for the ruthless elimination of Jewish competition. It is true that the foundation of new chain stores and consumers coöperatives was restricted, but this had the effect of creating a virtual monopoly in the hands of those already engaged in a certain trade. The small retail store faced the monopoly strength of larger stores which were protected against the entrance of new firms into the industry. Within the Estate of Trade and Industry, moreover, the department stores and chain stores carried more influence and control because they had a larger number of representatives. And on February 18, 1941, all remaining coöperative associations and affiliated organizations were dissolved.

THE ESTATE OF GERMAN HANDICRAFTS

The Estate of German Handicrafts, closely related to the organization of Industry and Trade, also indicates how far centralization of power to make and enforce decisions has been carried by the Nazis.[33] The "authoritarian" principle has been extended into the field comprising all firms and individuals entered in the Register of Craftsmen, as well as all journeymen and apprentices employed by them.[34] Every significant change in the preëxisting machinery of guild associations and handicraft chambers has been identical with the alterations tending to eliminate competition and promote the

[33] Act for the Preliminary Organization of Handicraft, November 29, 1933, and subsequent executive orders of June 15, 1934, January 18, 1935, and March 23, 1935.

[34] The predecessor of the present Estate of German Handicrafts was the German Handicrafts and Trade Chamber Assembly, divided regionally into 26 divisions, with 29 Handicraft Chamber and 8 Trade Chamber member bodies.

exercise of monopoly made under the National Economic Chamber.

The four main modifications introduced by the Nazis show clearly that policy-forming powers are being concentrated, systematized, and coördinated with other phases of economic and cultural activity. These changes were as follows: the system of Handicraft Masters (leadership principle) has been added, the regional organization promoted, and the coördinated new central body placed under the National Economic Chamber.

According to a decree of February 1939 all craftsmen suitable for employment elsewhere were struck off the Register of Craftsmen if the maintenance of their undertaking was not justified by economic reasons.[35] The maintenance of undertakings belonging to overcrowded branches of the handicrafts — e.g., those of the bakers, butchers, hairdressers, tailors, and shoemakers — was considered as having no economic justification. Questions of appeal from this decree come first to the Handicraft Chamber and in the second and final instance to the next higher administrative authority. Application for fresh registration may be made only after three years. The establishment of new undertakings is subject to the same regulation. The effect of this decree is to increase the supply of labor and at the same time eliminate competitors in crowded handicraft trades.

COÖRDINATION FOR SELF-SUFFICIENCY AND WAR

The military keynote of the Nazi economic experiment is beyond doubt. It found its extreme expression, outside of war, in the Second Four-Year Plan. Originally it was not intended to set up any special administrative machinery for the execution of the Second Four-Year Plan. General Goering, with a small body of expert advisers, was supposed merely to direct

[35] *Soziale Praxis*, March 15, 1939.

and coördinate the activities of the competent economic ministries and other government departments. An efficient war economy could not be run by the uncoördinated economic administrations which had grown up; hence state authority was centralized. The new organization was rapidly increased, notably at the expense of the Minister of Economic Affairs, who appeared rather unenthusiastic about the tremendous pace at which General Goering was carrying out the plan.

Constant duplication of and overlapping with the normal activities of the Ministry of Economic Affairs were removed by the drastic reorganization of the Ministry. General Goering was made supreme head of the economic organization of Germany for the purpose of carrying out the Second Four-Year Plan. Schacht gave up his post as acting Minister of Economic Affairs but remained for a while as president of the Reichsbank and received the rank of minister without portfolio, as personal economic adviser to Hitler. Walther Funk, former Assistant Minister of Propaganda, was appointed to Schacht's former post. Five departments were set up under the reorganization and were manned as follows:[36]

Industry and Raw Materials	Major-General of Air Force Loeb
Mining, Iron, Power	Major-General von Hanneken
Trade, Economic Organization, and Commerce	Herr Schmeer
Finance and Credits	Herr Lange
Foreign Trade, Foreign Exchange, and Exports	von Jawitz

There has been considerable speculation as to the meaning of this change in the central economic administrative machinery. The circumstantial evidence is against the opinion frequently expressed, that it was a victory of "left-wing" party elements. Funk himself is considered rather a "conservative" as far as economic policy is concerned. His speech at the time he was

[36] *News in Brief.* VI (March 14, 1938), 39.

given office on February 7, 1938, is very illuminating on this point: "The state guidance of economics must help the industrialist to keep his initiative unharmed by the superfluous official claims and bureaucratic trickery. The worst enemies of economics are the ignorants, the denouncers and — strange and paradoxical as it may seem to mention these in a Ministry — the bureaucrats. In the fight against them we must support economics." [37] Nor can Goering be considered either by tradition and temperament or by anything which he has said or done as representing the radical element in the National Socialist Party. He specifically discouraged the demands for the nationalization of war industries suggested in a campaign against armament profits carried on in the *Voelkischer Beobachter* during January 1938. On February 1, 1938, a Berlin dispatch announced: "Those close to General Goering deny that nationalization of heavy industry is contemplated.... Nationalization would only be a hindrance by bureaucratizing industry and killing the initiative of the industrialists." [38]

The more plausible interpretation of this reconstruction is one which sees its principal meaning in the greater influence of military elements in the administrative machinery. Soldiers and the old military bureaucracy took a leading role in the administration of the Four-Year Plan. Goering's closest collaborators were the "three colonels," one of whom, Loeb, as was shown above, was given a dominant position within the Four-Year Plan administration.[39] It is significant that the military men who shared in the direction of the war economy and the Four-Year Plan, although favoring strict control of industry in the interest of national defense, disapproved of the anti-capitalist campaigns and declared themselves unmistakably against all nationalization. For example, Colonel Thomas,

---

[37] *News in Brief*, VI, 39.
[38] *Le Temps*, February 2, 1938.
[39] *Le Temps*, April 3, 1938.

head of the war economy department of the Ministry of War, declared: "The execution is left as far as possible to private initiative. The German war economy will not socialize war industry. . . . The entrepreneur and the merchant should make money. That is what they are for." [40]

The military nature of the economy also received a most significant confirmation in another direction. In December 1937 General Goering, in his capacity of Reich Minister of Aviation, published a list of some twenty leading executives and engineers in the aviation industry who had been appointed "leaders of defense industries." A few days after the publication of these appointments, a statement was released to the press to the effect that a secret order regarding the creation of a special *Wehrwirtschaftsfuehrer-Korps* had been signed by Hitler as early as 1936. The official communiqué added that such leaders had already been appointed earlier for the army and navy (i.e., for the munitions and shipbuilding industries). The indications were that the German "leaders" would be responsible primarily for the secrecy of all plans and orders executed in industrial enterprises working for the army, navy, and air force. For this purpose they were required to take the same oath as army officers and were held responsible for speeding up work not only in their own enterprises but in their districts generally. Thus it appears that the increasing militarization of the economy has not wiped out the influence of the business man, for these leaders of defense were business men and technicians.

Toward the end of 1938 the government appointed "commissars" for the automobile, machine, and construction industries, and in January 1939 another was appointed for the power industry. The chief functions of these "commissars" were the promotion of rationalization and the extension of capacity, together with the determination of priorities for these industries.

[40] *Le Temps*, April 20, 1938.

In December 1938 Dr. Funk received a similar delegation of power for all of industry within Germany.

When war broke out, the Nazis needed only to readjust and tighten control measures. Government measures to insure coördination and centralize direction of economic activity, however, reduced the influence and independence of the semi-autonomous Estate for Industry and Trade. A Council for National Defense, headed by Goering, was introduced by Hitler on August 30, 1939.[41] The Council, endowed with sweeping power to coördinate business efforts, carried on its work by regional defense "commissars" whom Goering picked from the ranks of provincial governors and other high administrative officials.[42] Provincial food and lumber bureaus and district economic offices were set up, and the Estate was enjoined to coöperate with these offices as well as the central authorities.[43] In addition, the government appointed special commissioners to assist with the work of the Chamber of Commerce in the regional organization of the Estate. The task of these commissioners was to see that business concerns were properly provisioned with labor, transportation facilities, means of production, and power.[44] The government also regulated civilian consumption by means of newly introduced local food and economic offices. In December 1939 the "commissar" of machine production received additional and extensive powers to insure priority for the most urgent government orders.[45]

In order to increase the capacity of war industries and convert other factories for the production of war material, the National Service Law of September 1, 1939, enabled the government to conscript plants and raw materials of business

---

[41] *Voelkischer Beobachter*, August 31, 1939.
[42] *V. Beob.*, September 6, 1939.
[43] *V. Beob.*, September 8, 1939.
[44] *V. Beob.*, September 22, 1939.
[45] *Der deutsche Volkswirt*, September 22, 1939.

concerns. But along with the authorization to require business concerns to produce, stock, or surrender goods which the government might need, also went the provision that business should receive a "reasonable" return.[46] Under a decree of September 5, 1939, the Minister of Economic Affairs could require industrial enterprises to combine for the purpose of increasing efficiency, pooling patents, erecting new plants, and promoting exports.[47]

Centralization, however, reached its climax only in January 1940, when the administrative hierarchy was reorganized at the top to eliminate jurisdictional conflicts.[48] Goering, with the assistance of a council, was placed in command of all economic activities including the jurisdiction of private business, the war department, and government. Dr. Funk, as head of the Ministry of Economic Affairs and the Reichsbank, was made subordinate to Goering. Included in the council were the following officials: state secretaries representing the Ministries of Economic Affairs, Labor, Transportation, Interior, Reich Forest Office, and the Four-Year Plan; a delegate from the Nazi party; and the chief of the war economy office of the High Command.

Sharp curtailment of civilian consumption and shifts of industrial demand created by war inevitably bring losses and hardships. Instead of permitting the burden to rest on the particular industries undergoing losses, the Nazis introduced measures to distribute the cost of closing enterprises on the concerns which remained in business. By a decree of February 19, 1940, the associations under the aegis of the Reich Food Estate, the Chamber for Culture, and the Estate of Industry and Trade were directed to levy contributions on their members in order to defray the depreciation charges and inter-

[46] *D.V.*, December 22, 1939, p. 345.
[47] *Rgbl.*, 1939, I, 164.
[48] *Deutsche Bergwerkszeitung*, January 5, 1940.

est and rent payments of plants and shops which were compelled to close. The government also distributed its orders among as many factories as possible, particularly in depressed industries.[49]

The changes in organizational control thus went through several stages. The Nazis transformed the already highly organized entrepreneurial economy only in so far as was imperative to consolidate the political power of the party. To achieve this goal the Nazis sought the elimination of unemployment and the achievement of a strong military state.

During the first period they avoided a central bureaucracy for running business, and controls were relatively elastic. Executive responsibility was left to the leaders in the Estate of Industry and Trade and to the entrepreneurs. In the second period, when armaments were tremendously accelerated, a priority system determined the channels of production. When the war broke out, controls were expanded and tightened; commissioners were placed in control of vital industries in order to increase productivity and insure their operation. The entrepreneurial function was no longer risk-taking, for risks had been practically eliminated with the assurance of a market; this function was confined for the most part to organization and technical problems within the plant. Throughout Germany the entrepreneur was fashioned in the image of present-day corporate executives who manage but do not initiate new enterprises.

Expansion of business, rationalization, founding of new firms, and the introduction of new products were all determined by the prior interests of the war program. After the war program, and only after it, come profits.

[49] J. C. de Wilde, "Germany's Wartime Economy," *Foreign Policy Reports*, June 15, 1940, p. 88.

III

ORGANIZATION OF TRANSPORTATION

WHEN the Estate of Industry and Trade was set up early in 1934, transport formed a part of the Estate, but later, September, 25, 1935, it was considered expedient to establish a separate body because the government owned 80 per cent of all transport enterprises and the problem of control was considerably different from that in other areas of the economy. The extent

TABLE 2

RAILROAD MILEAGE

(*Not including the Saar*)

Year	Reich Railroad Mileage (*kilometers*)	Private Railroad Mileage (*kilometers*)
1927	52,404	3912
1932	53,730	4478
1936	53,894	4485

SOURCE: *Stat. Jahrb.*, 1929, 1934, 1938.

of publicly owned railroads — the only division of transport for which there are available figures — was little changed by the advent of the Nazis, as can be seen in Table 2.

In order to insure uniform administration, the Organization of Transport comprises all private transport undertakings, as well as those of the Reich, states, and municipalities. At the head of the organization is the Minister of Transport, assisted and advised by a National Transport Council which is made up of the following representatives:

(a) Concerns Responsible for Transport
  Leaders of the seven private transportation undertakings

Representatives from the Reich Railway Company, Post-office, and Air Traffic
General Inspector of Roads

(b) Transport Users
Two representatives of the Reich Food Estate
Six representatives of the Estate of Industry and Trade
One representative of the Chamber of Culture
One representative of the municipalities
One representative of the Labor Front

The Organization of Transport is similar in structure and functions to the Estate of Industry and Trade, with which it is connected through representatives in the National Economic Chamber. Structurally it is divided into seven functional bodies — sea-going shipping, inland shipping, motor transport, carrier services, rail vehicles, forwarding agencies, and auxiliary transport services — and fourteen regional and sub-regional divisions. It provides a medium for unifying services through each geographic area and machinery for effective division of traffic by types and regions. Under its general supervision, compacts are made for interchange of traffic at given points and at given rates.

Each organizational group or unit is run on the "leadership" principle and gives advice to members on trade subjects only, for determination of freight and passenger rates is left to the Minister of Transport and the National Transport Council. Advisory councils for various functional and regional subdivisions act as auditors. They must be consulted before the budget of any regional or functional group is drawn up. One meeting each year is held by each group, and at that time a resolution by secret ballot is submitted to determine whether the director of the group enjoys the confidence of members. Progress reports and financial statements are also read.[1]

The Minister of Transport not only dictates freight and

[1] Department of Overseas Trade, *Economic Conditions*, pp. 182–184.

ORGANIZATION OF TRANSPORTATION

passenger rates but also controls other important aspects of transportation — construction of equipment, schedules, and hiring of personnel. Control over such services is extremely important from a political point of view because it gives the National Socialists a chance to distribute rewards and spoils in the form of reduced rates to loyal party supporters.

TABLE 3
TRANSPORTATION

	1929	1932	1937
Rolling Stock			
Motor vehicles (1,000)			
Motorcycles	608	866	1,327
Automobiles	423	549	1,108
Motor omnibuses	11	12	17
Trucks and lorries	144	174	322
Railroads (1,000)			
Locomotives	26	24	24
Coaches	90	91	87
Freight cars	660	639	587
River boats			
With motor power	4,872	4,841	5,440
Without motor power	14,557	12,944	12,441
Traffic Carried			
Airplane			
Passengers carried (1,000)	97	99	323
Freight and mail carried (tons)	2,456	2,503	8,721
Railroads			
Passengers (million)	2,057	1,352	1,874
Freight (million tons)	531	307	547
River traffic			
Freight (1,000 tons)	140,669	73,744	133,080

SOURCE: *Stat. Jahrb.*, 1938, pp. 626 and 627, and I.f.K., *Weekly Report*, June 14, 1939, p. 65.

The effectiveness of this organization is difficult to measure because it is aimed at integrating and streamlining the transportation system. For example, more traffic may be carried by less rolling stock merely because transportation vehicles are being used at an excessive rate, or the same circumstance may reflect coördination of facilities. The most outstanding development presented in Table 3 is the growth in motor vehicles and air traffic. Although railroad freight traffic showed an increase over 1929, rolling stock declined.

THE VOLKSAUTO

The war called a halt to Hitler's plan for additional expansion of motor vehicles. As it was planned in 1938, a new automobile, the *Volksauto*, was to be produced under the auspices and responsibility of the German Labor Front and was to accomplish four purposes:

(1) Deflect the rising national purchasing power resulting from full employment, away from increased consumption
(2) Take up the slack in employment when rearmament was finally completed
(3) Gain enthusiastic support for the Nazis
(4) Capture the export market

Although Germany was the second biggest industrial country in the world, she was far behind other western powers in motorization. The number of automobiles had doubled in the first five years of Hitler's regime, but in 1938 there were still less than 1,500,000 pasenger cars, or one automobile for every 50 persons, as compared with one automobile for every five persons in the United States. Increased production which has largely been the result of a reduction in automobile taxes and a rise in national income was thus to be further stimulated by the ingenious *Volksauto* scheme.

The production of the car was planned at Fallersleben, near Braunschweig, where, according to Dr. Robert Ley, head of the Labor Front, was then being built not only the biggest automobile factory in the world but the biggest factory in the world of any kind. Toward its construction, the Labor Front, which had in 1938 an annual income of almost 400 million RM, had already advanced 50 million RM. The Labor Front also instituted a "pay-before-you-get-it" installment plan similar to that used by private building and loan associations in Germany. According to this installment plan, every would-be purchaser was to save five RM per week for the car, or sums in multiples of five, which were to be paid by means of purchasing a savings stamp and pasting it upon a card. When 750 marks had been paid, the buyer was to receive an order number entitling him to a car when the factory was able to deliver it. The total amount to be paid in amounted to 1,190 RM, not including delivery charges, and at the five-Reichsmark-a-week rate, payment stretched over about four and three-quarters years. Cash payments were excluded for domestic purposes. Cancellation of an order meant forfeiting of amounts already paid, except in special cases where 20 per cent might be retained. This installment prefinancing creates an enormous backlog of orders for after the war and at the same time forces additional saving during the war.

The *Volksauto* represents serious competition to private enterprise in the small-car market. Raw materials were reserved for its manufacture; all advertising and promotion costs were saved because the whole National Socialist regime, including the press, was enlisted in its promotion; and the dealer's profits was also saved because the "dealer" was the far-flung organization of the Labor Front.

SHIPPING

Private shipping also received special attention from the government. Since May 1933 the Nazis have accorded subsidies to German shipping as compensation for losses incurred in competition with foreign competitors. In addition to subsidies for operation, compensatory grants have been available to owners who have undergone losses as a result of being forced to sell foreign exchange to the Reichsbank. As a part of the public works program, 1933–35, German shipping companies which placed contracts with German yards for reconditioning, replacements, and new constructions could obtain a government grant to the extent of 30 per cent of the value of the order. A vigorous national campaign supporting the use of the German flag by shippers brought increased business to German shipping and to other nationalities using German ports — mainly at the expense of tonnage under the British flag.

The effect of government assistance to private shipping is clearly reflected in the increased freight carried by German ships, as well as in the improvement in ship capacity since the depression. Statistics for private German shipping are presented in Table 4.

TABLE 4
PRIVATE GERMAN SHIPPING

	1929	1932	1937
Ships	3,939	3,590	3,668
Motor ships	2,087	1,890	2,032
Gross tonnage of ships (1,000 GT)	4,242	3,957	4,132
Freight (1,000 tons)	48,480	33,317	56,695

SOURCE: *Stat. Jahrb.*, 1938.

The government, moreover, assisted the financial reorganization of the Hamburg-America Line and the North German Lloyd. In 1930 these two companies amalgamated under a fifty-year agreement which resulted in approximately 70 per cent of the total German tonnage coming under the centralized control of the so-called Hapag-Lloyd Union. Unfortunately this union did not meet the expectations of its financial sponsors, and reorganization was instigated by the government very soon after the succession to power of the Nazis. A new Union agreement of February 1, 1935, was made for a period of fifteen years.

Several blocks of stock were severed from the Union as early as 1934 — the Hamburg–South America Steamship Company, the German Africa Lines, and the German Levant Service. Companies financed with private capital were formed in Hamburg and Bremen to acquire the tonnage and other assets from the Hapag and Lloyd. These two companies now operate through a holding company, for which purpose the old German Levant Line of Hamburg has been retained. The two companies were heavily indebted to the government, but as a result of reorganization a large part of this debt was wiped out.[2] Shipping tonnage after the reorganization was divided as follows:

	Number of Vessels	Gross Tons
Hamburg-America Line	111	743,112
North German Lloyd	114	589,954
Hamburg-South America S.S. Co.	36	279,260
Hansa Line	37	262,165
German Africa Lines	30	138,658

So far as the elimination of competition is concerned, the new agreement resembles the old one; the two companies have agreed to coöperate in regard to sailing lists and vessels to be used on various routes, as well as in the methods to be adopted in canvassing for passengers and freight. Provision has also been made for coöperation between the companies in protecting the seaport interest of Hamburg and Bremen. In the interests of more flexibility in management and increased sense of responsibility, joint operation on a profit-pooling basis has been restricted to certain vital routes — the east and west coast of Central and South America, Cuba, Mexico, and the Far East.

The results of this arrangement have been that the Hapag acquired increased interests in the trade with the Americas, while the Lloyd obtained greater importance in business with the Far East, a route on which it operates three 18,000-ton fleet liners built by the government.

[2] Department of Overseas Trade, *op. cit.*, p. 185.

SPECIAL WAR MEASURES

Even before the war the government had been much concerned about the bottleneck created by inadequate railroad facilities. In an effort to increase efficiency with the outbreak of war, special commissioners were appointed for each railway district to establish priorities. According to another decree, passenger trains were drastically reduced and freight cars were required to be unloaded immediately, even though the destination was reached on Sundays or holidays.

In addition, the government reduced its road building by half, while continuing the remainder only on sections of great strategic importance. In a six-year period the Nazis had completed 3,065 kilometers, partially completed 1,383 kilometers, and undertaken preliminary work on 2,449 kilometers.[3] The volume of motor traffic also had to be decreased in order to save gasoline, and to this end the use of motor cars for private purposes was banned.[4] Even for public purposes, only small cars were to be used. From the beginning of 1940 long-distance trucking was prohibited unless recognized as essential to the prosecution of the war.

[3] Deutsche Bank, *Wirtschaftliche Mitteilungen*, January 1941, p. 3.
[4] *D.V.*, September 29, 1939.

IV

CORPORATIONS

LARGE business as represented by the German corporation is still an important force in German economic life since Hitler has retained the outward trappings of a pre-Nazi economy. Even war controls were not radical because Germany had had a long period of war gestation and achieved military victories early. Industry had six years in which to re-tool before war started, and thus the machine tool industry did not receive the impact of war all at once. A similar situation existed in other industries. Moreover, capacity created during the twenties, encouraged by American loans, was excessive for peace-time needs in a world marked off by protective tariff barriers. As early as 1933 Hitler put this unused capacity to work in preparation for war. When war actually broke out, there were enormous inventory reserves, and extension of plant was not quite the problem that it was in Great Britain or the United States.

THE CORPORATION LAW OF 1937

Despite the fact that during the early days of the Nazi movement certain party members vehemently denounced the corporation and the large firms, four years went by before any general law was passed specifically covering the corporate or limited liability form. This long delay is undoubtedly a reflection of lack of unanimity between conservative and radical forces; and when the law was finally promulgated its changes were not radical.

After party leaders decreed that the plank in the Nazi platform which called for nationalization of trusts would not be carried out, the more radical members attempted to do away with the corporation altogether. They proposed to place all business upon a partnership basis, giving as their reasons for such action the lack of personal relations within the "anonymous" corporations and the various abuses of the privileges granted under the corporate form. A law of July 5, 1934, was apparently aimed as a sop to these demands, but as it turned out the law did not result in any large changes in the total capital stock in corporations. According to this act, a corporation under certain specified conditions might change to another form of organization without a majority vote of the stockholders which had previously been necessary.[1] It thus facilitated mergers and nationalization as well as a change to the unlimited liability form. As a result of this law, after four and one-half years of operations, from July 1934 to December 1938, the capital involved in a shift from corporate form to single firms or partnerships came to 585 million RM, or 4 per cent of the total capital stock of all corporations in 1938, capital in mergers to 664 million RM, and capital in nationalizations to 309 million RM. The law thus had a greater influence in encouraging mergers than it did in promoting the unlimited liability form; only very small firms gave up limited liability privileges. Sixty per cent of the total capital involved in a change from corporate form to individual proprietorships or partnerships took place in the finishing industry. The commercial trades industry had the second largest share of such changes — approximately 10 per cent of the total capital transferred to the unlimited liability form of business organization.[2]

A law of October 9, 1934, was also introduced to appease certain radical demands but, like the earlier law in July 1934, proved to be of little importance. It relaxed the rules and regu-

[1] *Rgbl.*, 1934, I, 569.
[2] *Vjh. Stat. d. D. R.*, 1939, I, 116.

lations governing bankruptcy procedures for joint stock companies in order presumably to encourage a decline in this form of business organization. During a four and one-half year period the October 1934 act affected only 240 companies with a total capital of 51 million RM.[3]

The corporation law of 1937, however, had teeth in it and forced smaller firms to give up the corporate form. According to this act, companies with a capital stock of less than 100,000 RM must be dissolved, merged with larger companies, or must change to the unlimited liability form. This provision automatically meant the elimination of about 20 per cent of the existing companies, although only 0.3 per cent of total corporate capital. The provisions under the 1937 law in regard to new firms are even more demanding. No new firm may acquire the limited liability form unless its capital stock totals at least 500,000 RM. By decreasing the number of corporations, the law has facilitated regulation but at the same time tends to restrain competition and protect the interest of large companies. It also makes it more difficult for individuals in the middle income classes to avoid the higher tax rates on individual income by means of setting up a family corporation which is obliged only to pay the lower corporate tax rates. Individuals with very large incomes, however, are not affected by the minimum requirements.

Speculation in stocks by the small saver is made practically impossible by the provision that no shares may be issued with a par value of less than 1,000 RM.

In addition to prohibiting the use of the corporate form for small firms and discouraging the wide distribution of stock ownership, the corporation law of 1937 attempts to centralize control within the corporation and promote the objectives of the National Socialist state. It also makes certain gestures toward meeting the demands of certain party members for minor re-

[3] *Ibid.*

forms. The old law, the Civil Code of 1884, covered individual proprietorships as well as corporations. Nazi regulation represents a separate law for the limited liability form.

Centralization of Control — Under the Civil Law Code, the corporation had three organs of control. First there was the *Vorstand*, a board of managers similar to a board of directors in that it did much initiatory work. Rarely were the members of this board capitalists, but rather highly paid "civil servants" of industry. Departmental heads were frequently placed on this board of managers who might also be called the administration. In many cases the decisive power of the undertaking ultimately rested here; in other cases it lay in the second statutory body, the *Aufsichtsrat*, or supervisory board. This was generally the seat of the representatives of the largest holders of capital. Some trusts found their real leader in the chairman of this supervisory board. In contrast to this principle, however, Emil Rathenau and Deutsch, the leaders of Germany's gigantic electricity trust, were chief managing directors without votes in the *Aufsichtsrat*. The third body was the general meeting of the stockholders.

Under the present act the sole charge of the business has been placed in the hands of the board of managers selected by the board of directors. Control can be centralized further by the clause which enables a board of managers to consist of one man and to place all responsibility of the board of directors in the hands of its chairman. The board of directors for the largest companies has been decreased from a maximum of 30 to 20 members and proportionately less for smaller companies. No single director may hold more than 10 directorships (formerly 20). And the prohibition of the Civil Code against membership in both the board of managers and directors of any one company has been retained.

The term of office for a member of the board of managers is five years and is renewable. A member may be replaced only if he is grossly negligent of duty or is unable to attend to ordinary business transactions. The earlier proposals for complete liability of managers to the stockholders were not incorporated in the act. Managers are required to exercise the care of an ordinary and conscientious head of a business, but cannot be held responsible for changes due to the business cycle, or losses beyond the control of the management. Nor are they liable for operations carried out by the specific order of stockholders. Outside of these exceptions, managers must repay stockholders for losses.

The stockholders presumably exert an indirect control over the board of managers through their selection of the board of directors. Actually their influence amounts to very little, for, if numerous, they may be widely scattered, with conflicting interests, and hence difficult to organize for the purpose of exerting any authority. In Germany, as is true with corporate organization in other countries, various factors have encouraged separation of ownership and control — lack of accurate knowledge, financial independence and experience of stockholders in dealing with the many complex activities of corporations, as well as the numerous devices for controlling stockholders' votes.

Under the old law, shareholders had the right to decide final distribution of net profits. The present act maintains this right, but it is of little significance, since management may still determine deductions for depreciation before arriving at net profits. Moreover, the Loan Fund Act of 1934 requires that declared demands over 6 per cent (in certain cases 8 per cent) must be invested in government bonds. Within these limitations and the restrictions set forth under contracts to bondholders and preferred shareholders, the stockholders may decide on the final distribution of net profits.

The interests of governmental bodies as stockholders are presumably to be protected by their right to send their own special representatives to the board of directors. But in view of the fact that the number of representatives may not exceed one-third of the directors, the influence and importance of such representatives could hardly be decisive. As under the old law, stockholders owning at least one-twentieth of the capital stock may call a meeting for good cause. Preferred stockholders are to be given voting rights if the dividends are not paid in full for two consecutive years; the importance of this provision as a protection to the preferred holders obviously depends upon the relative amounts of common and preferred stock. Also, as in the Civil Code law, the independent auditors licensed by the state and chosen by the stockholders go over the company's books annually.

An increase or reduction of capital is possible only after consent of three-fourths of the stockholders; a similar provision was contained in the old law. Under the 1937 act, however, management can avoid the requirement that all increases in capital must be made with the specific permission of the stockholders by having a clause in the charter permitting blanket increases. The new law extends the previous limitations on loans to officers. As in the earlier code, loans or credits to officers are permitted only with the consent of the board of directors. At present the officials to whom this limitation applies have been increased to include those having the right to hire and fire other employees, as well as the board of managers.

None of the provisions named could be called very revolutionary. The regulations are less stringent than those of the Securities Exchange Act for the United States. Although centralized control has been facilitated, the management and responsibility are unchanged.

Publicity and Correction of Specific Abuses — The provisions for "personalizing" the corporation hardly seem sufficient to cast much light on the individuals who actually control a company. The requirement that the names of all the members of the board of managers and the chairman of the board of directors must appear on all business letters and financial reports is extremely limited in its effectiveness. Stockholders have the right to request any information they desire unless it will injure competitive standing or be against the interests of the general welfare. This provision of the law leaves room for extremely broad interpretation. A board of managers would not have much difficulty in finding reason to refuse information under such loosely stated provisions as competitive standing and general welfare.

Connections of the firm with combines, cartels, and similar organizations must be revealed in annual reports to the stockholders. This facilitates government regulation, but still leaves the stockholders uninformed as to the individuals holding the largest blocks of voting stocks and hence playing the largest role in control.

The provisions covering proxy voting are also mild. The voter must indicate which shares belong to him and which to others; banks are required to secure permission of the owner before using the voting power of shares deposited in their care by customers; and no subsidiary company may vote shares of the parent company which it may own. The first two rules serve merely to broaden the former requirement that all those voting at meetings of the stockholders, together with the amount of their holdings, must be listed.

Incorporation may be refused by the courts if the promoters have not observed the letter of the law, if the preliminary audit is unsatisfactory, or if goods and services taken in exchange for capital stock have been overvalued. All expenses in operation and promotion — including salaries, participations in profits, commissions — of managers and directors must represent a "reasonable" reward for services rendered, as well as have a "reasonable" relationship to the conditions of the company.

Control for State Aims — The most far-reaching regulations are those which provide for direct state intervention in cases where a company endangers the general welfare through conduct unbecoming to the corporation law or "contrary to the principles of responsible business management." Dissolution by the Reich Economic Court upon application of the Minister of Economic Affairs awaits the firm guilty of such conduct. The language used is in such vague terms that the exact meaning depends upon the interpretation of the courts and the Minister of Economic Affairs.

The Corporation Law of 1937, like other institutional changes, could develop in almost any direction. There is the possibility of widespread state intervention, but at the same time the impetus given to centralized control might strengthen the power of business management over employees and stockholders. It seems aimed to insure fair practice in the business community as well as to increase centralization of control in order to make business commands more efficient. At the same time, however, it guarantees the primacy of the state. All three ends were consistent within the war economy.

CORPORATE PROFITS AND DIVIDENDS

Evidences of corporate experience and policy, such as are ordinarily drawn from examinations of balance sheets and income accounts, are among the significant details in any satisfactory picture of the Nazi armament boom, its background, and its sequel. The following analysis presents and interprets an historical record, 1926–39, of changes in profit ratios for each of several important industrial classes, and for all combined. Special attention is devoted to industrial groups designated as heavy industry, light industry and luxuries, as well as to certain selected classes, such as banking, chemicals, transportation, utilities, and building trades and materials.

Profit ratios were computed by dividing corporate net income by net worth. Data for net worth and net income are published annually by the German Statistical Office [4] and the statistical coverage is remarkably good, amounting roughly to 92 per cent of total corporate capital and 46 per cent of the total number of corporations. The various items in the balance sheets and in the profit and loss account, including deductions for depreciation and obsolescence, are in general well defined by legal regulations governing accounting procedure. It must be remembered that the course of profit rates shown here tells nothing concerning the small business man who is not incorporated, and omits any report on very small corporations — the 54 per cent which represent but 8 per cent of total corporate capital.

Recovery measures, based as they have been on large national works and rearmament, have inevitably favored big firms, which alone were in a position to carry out large scale orders. This is true, despite the fact that the protection of the small man against the big constituted one of the most important points in the National Socialist economic program. In addition, small enterprises, being both small and numerous, had difficulty in obtaining raw materials rationed under the Second Four-Year Plan. Technical changes in methods required by changes in the types of raw material have always been more difficult for the smaller firms. The smallness of the original or basic quotas has also tended to handicap the new and growing firms.

The development of the profit ratios for 1926 through 1938 can be seen in Chart I, which is based on the figures presented in Table 5. The average ratio for all industries reveals a substantial improvement as early as 1932, but this improvement merely means a smaller aggregate loss for the group than in

[4] For detailed discussion of the goodness of the statistical material and method used in computing profit ratios, see Maxine Sweezy, "German Corporate Profits: 1926–38," *Quarterly Journal of Economics*, May 1940, pp. 384–398.

TABLE 5
Profit Ratios of German Corporations According to Industrial Classes and Composite Groups

(Unit: Per cent — net earnings divided by net worth)

	1926	1927	1928	1929	1930	1931	1932	1933	1934	1935	1936	1937*	1938*
I. Heavy industries, of which													
Mining	5.86	6.50	6.97	7.81	5.46	1.81	2.84	.56	2.13	4.03	4.55	5.66	4.82
Iron and metal extraction	4.71	4.57	3.64	3.56	2.50	−11.64	.20	−.76	5.98	5.25	5.89	6.53	3.67
Iron, steel and metal manufacturing	−4.30	5.16	3.86	1.65	−4.83	−21.32	−5.55	.18	−.95	3.13	4.94	8.57	5.04
Machinery and apparatus	−1.34	4.17	4.26	2.68	.43	−13.79	−6.53	−.73	2.72	4.37	7.64	7.88	7.56
Conveyances	.51	.82	3.57	−2.98	−10.70	−13.28	−14.11	−4.50	9.64	12.15	12.31	9.01	9.84
Electrical industry	4.12	6.65	6.82	7.09	1.75	−16.11	−13.84	.20	5.93	5.68	8.47	9.75	7.91
Rubber and asbestos	.50	6.38	5.99	5.61	1.28	−11.09	−5.54	4.18	7.10	7.59	8.89	10.48	8.95
Average for group	2.10	4.89	5.02	4.06	1.63	−12.60	−6.94	1.59	4.92	6.01	7.53	7.41	6.44
II. Light industries, of which													
Paper	6.26	8.97	8.70	6.92	3.34	−10.14	−9.87	−5.68	4.42	3.78	6.34	7.89	5.97
Paper finishing	−.49	5.39	5.94	2.97	1.12	−7.62	−2.81	1.43	−1.01	1.81	6.07	3.85	2.44
Leather and linoleum	5.36	9.76	1.40	−.20	2.37	−6.48	1.53	−13.19	7.91	5.47	7.91	8.85	7.14
Prepared foodstuffs	4.95	5.63	5.04	4.94	4.61	−1.36	2.22	2.22	4.30	5.38	5.33	....	5.67
Clothing	7.97	12.67	10.65	8.08	3.81	−5.52	1.80	6.15	7.32	6.11	7.26	7.77	8.18
Fishing	5.65	4.92	1.78	8.95	−6.56	−16.61	−12.15	5.65	5.52	9.82	6.51	....	....
Textiles	3.75	9.32	4.57	−6.93	−3.76	−9.16	−7.77	1.99	4.86	4.06	5.23	7.34	7.49
Commerce	4.11	5.79	4.84	3.47	1.63	−4.93	.24	1.61	2.71	3.52	5.60	4.58	5.26
Average for group	4.25	7.81	5.37	4.42	2.11	−7.73	−4.08	2.38	4.63	4.99	6.16	6.71	6.02
III. Fine mechanics and optics	....	3.99	3.79	.84	−5.90	−20.77	−6.34	−.26	3.42	3.91	7.40	....	....

IV. Luxuries, of which										
Wood and wood carving	.92	4.28	2.55	−2.18	−8.45	−26.83	−18.20	−1.15	6.93	3.97	5.32	2.51	3.88
Musical instruments	6.76	11.89	9.99	9.41	−1.31	−26.66	−26.48	−11.77	−16.96	−1.81	.53	−.20	−1.90
Restaurants, drinking, inns	2.68	2.92	3.55	1.84	1.20	−5.02	−10.45	−11.80	−16.71	−2.59	3.01	3.00	2.09
Reproduction (films)	5.31	5.92	3.84	3.43	−.30	−12.07	−8.52	−5.68	−3.61	−2.07	.48	4.05	4.87
Average for group	3.91	6.25	4.98	3.12	−2.06	−17.64	−15.91	−7.60	−7.59	−.62	2.09	2.04	2.02
V. Building trades and materials	3.48	9.02	8.58	3.59	4.59	−12.61	−11.84	1.07	3.72	3.25	6.72	7.60	9.06
VI. Chemicals, of which	5.26	8.58	9.45	7.71	6.07	1.39	2.02	4.10	5.23	5.33	5.99	6.21	5.88
Misc. including I.G. Farben	8.20	10.19	11.15	9.59	9.52	4.62	5.19	5.43	5.56	5.34	6.07	6.17	6.10
VII. Transportation (Railroads and buses not included)	4.21	4.89	3.94	4.52	2.42	−18.48	−14.29	−.14	−.34	−.06	.53	1.77	2.49
VIII. Water, gas, and electricity	6.94	6.60	6.38	6.28	5.58	4.57	4.16	−.44	4.56	4.82	5.43	5.46	4.95
IX. Banks, Insurance Companies and financial houses, of which	10.44	8.69	8.85	8.55	4.57	−32.25	−5.45	2.01	3.92	4.22	4.53	6.28	5.76
Mortgage banks	10.58	9.66	10.45	10.32	9.92	−5.19	4.16	3.46	3.45	3.61	3.78	3.61	3.38
Investment companies	−3.43	11.15	7.94	−14.84	8.54	−2.67	3.19	1.52	5.25	4.18	7.79	6.63	7.57
Financial holding companies	3.33	3.73	6.53	4.88	1.47	−10.92	−2.70	1.20	2.77	3.26	6.23	5.86	6.75
Insurance companies	9.42	11.11	11.22	6.72	9.76	−8.51	8.76	10.06	11.04	9.74	9.36	9.42	8.30
(No division for large commercial banks)													
Average for all industries	5.17	6.81	6.18	4.53	2.77	−10.59	−3.65	.74	3.52	4.22	5.30	6.16	5.66
Same after deductions for estimated corporation taxes †	4.13	5.45	4.94	3.62	2.22	−10.60	−3.66	.59	2.82	3.38	3.97	4.31	3.68

* Preliminary figures.
† Although the profit ratio are negative in 1931 and 1932, some firms made profits and hence paid taxes.
SOURCE: Computed from data in *Vjh. Stat. d. D. R.* for the years 1926-32; *Stat. d. D. R.*, Bands 493 and 525, and *W. und S.* (1939), XIX, 596-597.

1931, which was marked by the all-time low of −10.6 per cent. Profits appear for the first time in 1933 and show an increase until 1938, when profit ratios began to slump somewhat. The decline in the rate of increase after 1934 is explained principally by the failure of the light or consumers' goods industry to keep

CHART I

CORPORATE PROFITS, AVERAGE FOR ALL INDUSTRIES, 1926–38

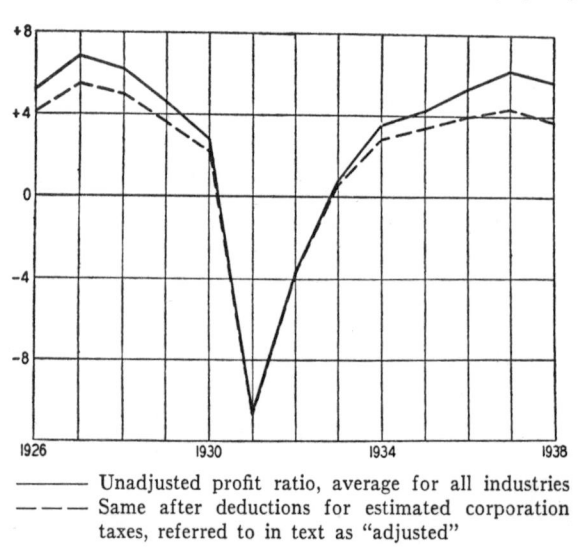

——— Unadjusted profit ratio, average for all industries
— — — Same after deductions for estimated corporation taxes, referred to in text as "adjusted"

up with the improvement being made by heavy industry. The fact that the largest profit for all industries during the Hitler regime — 6.16 per cent in 1937 — fell short of the highest point in the previous boom can be accounted for by the lag in the consumers' goods group and by the failure of luxuries to recover.

The configuration of Chart I is the result of the various measures incorporated under the two Nazi Four-Year Plans. The First Four-Year Plan raised national income and the level of demand; the second plan continued to reinforce the level of private demand with state orders. The government's general policy of fixing wage rates and forcing reductions in interest rates reduced costs. True, prices were also stabilized, but an increased rate of profit was to be expected with increased utilization of plant.

CHART II
CORPORATION PROFITS FOR SELECTED INDUSTRIAL GROUPS, 1926-38

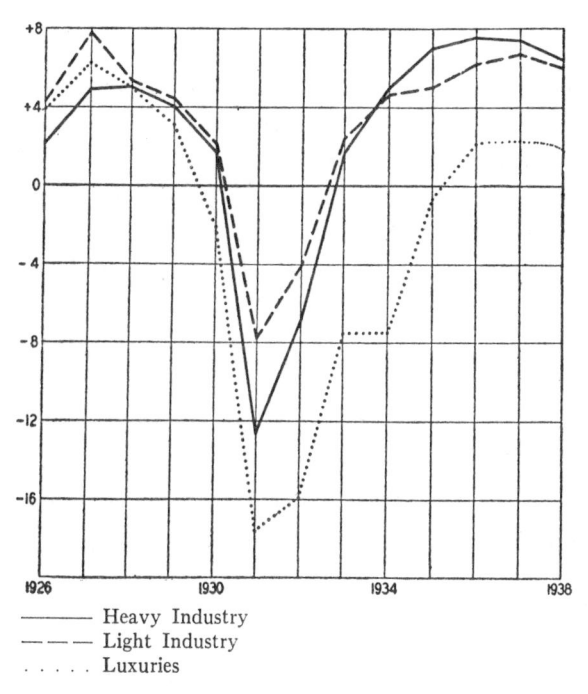

——— Heavy Industry
— — — Light Industry
. . . . . Luxuries

A decline in write-offs or expenditures for depreciated equipment and factory made profits appear unusually large for the period, 1933-37.[5] This decline by itself could never have brought the significant improvement, but, in so far as profits

[5] *Stat. d. D. R.* (1938), Band 525, p. 91.

are a reflection merely of decreased expenditure for upkeep, they are spurious.

An investigation in greater detail of the various industrial groups presented in Table 5 and Chart II reveals that certain industries gained much more than others in the general revival of profits. It is strikingly evident that those industries which suffered losses greater than the average in 1931 were, with two or three exceptions, the same industries which made profits greater than the average in 1936–37. This, however, is to be expected on the ground that profits in these industries are most sensitive to the cycle, falling to the lowest depths in the slump and rising to pinnacles in prosperity. Another observable feature of these ratios is not so platitudinous. Those industries which were worse off in 1931, with four exceptions, had profit ratios greater in 1936–37 than at the peak of the pre-Nazi prosperity. This can be seen clearly in the list presented below.

Losses Greater than Average for 1931	Profit Rates Larger in 1936–37 than in 1927–28
PRODUCERS' GOODS INDUSTRIES	
Iron and metal extraction	Iron and metal extraction
Iron, steel and metal mfg.	Iron, steel and metal mfg.
Machinery and apparatus	Machinery and apparatus
Conveyances	Conveyances
Ship building	Ship building
Electrical industry	Electrical industry
Rubber and asbestos	Rubber and asbestos
LUXURIES	
Wood and woodcarving	Wood and woodcarving
Musical instruments	....
MISCELLANEOUS	
Fine mechanics and optics	Fine mechanics and optics
Building trades and materials	....
Transportation	....
Fishing (Light Industry)	....

Although the significance of this may not be commonplace, the results are to be expected in the type of "recovery" experienced in Germany under the Hitler government. It means simply that during an armament boom the producers' goods industry experiences extraordinary increases in profits, while the earnings on consumers' goods, and particularly the luxury industries, remains below the level for a normal prosperity. Whereas the profit rates in the light industries were above the rates for heavy industry during the pre-Nazi prosperity and depression, after 1933 the relative position of these two industrial classifications was reversed. During the entire period of the Nazi regime, profit ratios for heavy industry were above the level for the ratios of light industry.

Another characteristic of the Nazi boom is evident in the movement of the profit rates of the luxuries group. This series is generally very sensitive to cyclical movements both up and down, but during the Nazi boom remained considerably below the average rate of profits for all industries. In the pre-Nazi period of flourishing business conditions, the profit rate of this group was above that for heavy industry, also strongly susceptible to booms and slumps, but the Nazi armament program sent the profit rate for heavy industry above that for luxuries.

Profits rates in the musical instruments industry has never revived under the Nazi type of recovery although those for the wood and woodcarving industry, also classified as luxuries, showed considerable improvement, being influenced by the government's attempts to increase the production of commodities which can obtain foreign exchange. Profit rates in the fine mechanics and optics also showed an upward tendency, for companies producing lenses, measuring instruments, etc., used in war equipment were included under this particular classification.

Attention will now be devoted to the remaining series which were omitted from Chart II in order to avoid confusion. The building trades and materials group has a profit rate above the average for all industries. The increase since 1936 not only in the profit rates themselves but also in the rate of increase of these ratios coincides with the expansion of industrial building for *Ersatz* and armament factories. From 1932 to 1938 the extent of public building construction increased enormously, with government orders amounting to about 80 per cent of the estimated 11½ billion RM spent in construction. Because of the demands of economic and military armament, the building industry has had little of its resources available for private dwellings.

The chemicals group, along with mining, was the only industrial classification as presented here which weathered the depression without losses. The chemicals series, however, which must be strongly affected by military production, is still below the previous boom. The miscellaneous class which the official figures show under chemicals is heavily weighted with the I. G. Farbenindustrie, A. G., the largest corporation in Germany, and stands above the ratios for the whole group. Private transportation, excluding railroads and buses, on the other hand, was very slow in recovering, with a profit ratio still below 1 per cent in 1936. The continued depression in this group may be explained by the emphasis which the Nazis have placed on the development of the *Volksauto*.

The ratios for water, gas, and electricity — like chemicals and mining — stood up fairly well during the depression but made losses in 1933. When the Nazis came to power, considerable disorganization was engendered in public utilities. In certain important instances, the national and state governments as well as a number of municipalities retreated under favorable conditions from key positions and ownership in this industry. Reasons other than the general disorganization probably exist, but they are not evident, and the return to private industry did not take place until 1935.

CORPORATIONS

Banks, insurance companies, and financial houses showed the greatest changes of all during the 1932–37 period, but the profit ratio in the last year was only 6.28 per cent as compared with 8.85 per cent in 1928. The change was so great because this group made losses amounting to 32.25 per cent of net worth, the largest "loss rate" of any industry. Unfortunately the statistical material does not include a separate tabulation for the commercial banks which suffered most heavily in the bank crash of 1931. From the separate classifications which have been made, it can be seen that profit rates for mortgage banks have had the mildest recovery. The profit rates of the investment companies and the financial holding companies are about even with those in 1928. The ratio for insurance companies was practically up to the previous prosperity level in 1934 but has declined slightly since that time. This whole group is enormously influenced by changes in government financing policy. In spite of the prodigious extent of government operations on the money and capital markets, the private banking industry has made fair profits. The increase of the profit ratio for investment companies of more than 3 per cent in 1936 is explained by the fact that private issues of corporations increased considerably in that year.

The provisional statistics reveal that during 1938 the Second Four-Year Plan had the effect for the first time of lowering the profit ratio of heavy industry. Several factors were involved in this decrease. Heavy industry, which was facing diminishing returns because of fully utilized capacity, was permitted to expand plants for armament production or build new ones in enterprises where profits were guaranteed by the state, thus expanding net worth. Labor output per man hour has undoubtedly decreased as a result of longer hours. A twelve-hour day has been reported in many factories. Under the pressure of armament orders of the last four years, the wear and tear on factories and equipment was probably excessive, and adequate provisions for maintenance may have been omitted. An indication of this is seen in the official report that there was an abnormal increase during 1938 in the depreciation accounts for some of the subdivision groups included under heavy industry.[6]

The improvement in the profit ratio for light industry during

TABLE 6
DIVIDEND PAYMENTS

Year	Dividend Payments as Per Cent of Share Capital, All Corporations	Dividend Payments as Per Cent of Share Capital of Dividend-Paying Companies	Percentage of Total Corporations Paying Dividends
1927/28	6.7	8.5	80
1928/29	6.4	8.4	76
1929/30	6.5	8.5	76
1930/31	4.8	7.2	65
1931/32	2.6	6.6	39
1932/33	2.9	6.2	45
1933/34	3.1	6.0	52
1934/35	3.6	5.9	61
1935/36	4.2	5.7	73
1936/37	4.7	6.1	77
1937/38	5.2	6.5	80
1938/39	5.6	6.6	81

SOURCE: *Stat. d. D. R.*, Bands 493 and 525, and *W. und S.* (1939), XIX, p. 595.

1937 may be attributed to orders for military uniforms and similar commodities which were placed with factories not yet working to full capacity. The increased demand brought about by the *Ersatz* activity, which entails particularly the encouragement of synthetic food and clothing production, was partially offset by the decline in production based on natural raw materials. The demand for peace-time consumers' goods in 1938 remained considerably short of the pre-Nazi recovery, for the orphan child of Nazism has been the consumer. His consumption has been limited by government propaganda and policies so as to divert buying power into capital investment.

[6] *W. und S.*, XIX (1939), 595.

Dividends — Despite a growth in undivided profits, earnings rose so rapidly that dividend payments have also shown a steady upward course since 1932. Variations in dividend payments from year to year are presented in Table 6. The last year, 1938–39, for which a breakdown by industrial groups is available reveals that dividends of more than 6 per cent were paid in the rubber and asbestos industry (7.6 per cent), mining (6.9 per cent), chemicals (6.8 per cent), ferrous metal manu-

TABLE 7
AVERAGE SIZE OF GERMAN CORPORATIONS *

	\multicolumn{3}{c}{CAPITAL STOCK IN 1,000 RM}
	1928/29	1938/39	1939/40
Producers' goods	3,120	6,592	6,874
Consumers' goods	1,290	1,675	1,845
Luxuries	639	993	?
Building trades and materials	1,062	1,375	1,565
Chemicals	3,791	7,579	8,380
Transportation (private)	4,138	3,076	?
Water, gas, and electricity	7,302	12,532	13,446
Banks	2,932	5,405	6,326
Financial holding companies	8,409	12,070	?
Insurance companies	2,078	3,209	3,219
All corporations	1,958	3,397	3,798

* Averages for various industries were computed by dividing capital stock for a given industry by the number of corporations for that industry. The average for all corporations was obtained by dividing the total capital stock by the total number of corporations. *Stat. Jahrb.*, 1929, p. 443; *Vjh. Stat. d. D. R.*, 1939, I, 116; Deutsche Bank, *Wirtschaftliche Mitteilungen*, November 30, 1940.

facturing (6.6 per cent), and insurance companies (11.17 per cent).

BANKRUPTCIES, NEW FIRMS, AND MERGERS

The increasing importance of big business and heavy industry is clearly demonstrated in statistics on the average size of corporations, bankruptcies, and new firms. According to figures presented in Table 7 the average size of German corporations as measured by amount of capital stock was approximately 95 per cent greater in 1939–40 than in 1928. This increase may be attributed to the fact that producers' goods, normally large, play a more important role in a war economy. But not only did producers' goods corporations play a more important role, their average size was also more than twice as large in the war economy as it had been in the pre-Nazi prosperity (see Table 7).

The advantages which accrued to large firms in producers' goods industries are also revealed in statistics on bankruptcies and the establishment of new firms. There was a striking decline in all bankruptcies, but particularly so for large corporations in heavy industry.[7] At the same time the average size of new firms, as measured by capital stock, increased enormously; and during six years, 1933–38, of the Nazi regime 61 per cent of the total capital stock in new firms went into heavy industry.[8]

[7] *Stat. Jahrb.* for various years and *Vjh. Stat. d. D. R.*, 1939, I, 115.

DECLINE IN BANKRUPTCIES

	Number of Corporations	Capital Stock (*1,000 RM*)		Number of Corporations	Capital Stock (*1,000 RM*)
1926	260	63,200	1933	65	33,600
1927	90	32,200	1934	41	14,900
1928	76	24,000	1935	24	9,400
1929	116	37,900	1936	19	2,800
1930	129	69,200	1937	8	2,000
1931	201	204,000	1938	7	1,900
1932	134	84,500			

BANKRUPTCIES BY INDUSTRIAL GROUPS
(*Per cent of total capital stock in bankruptcies*)

	1928–32	1933–38
Heavy industry	39.6	36.2
Consumers' goods and luxuries	53.6	59.1
Banks, etc.	6.6	4.7
	100.0	100.0

Computed from data given in *Stat. Jahrb.* for various years and *Vjh. Stat. d. D. R.*, 1939, I, 115.

[8] The average size of new firms as measured by capital stock amounted to less than 1 million RM for the pre-Nazi prosperity but had risen to almost 10 million RM in 1939 (*Stat. Jahrb.*, 1934, p. 365; Deutsche Bank,

Evidence on mergers completes the general picture. Not only was the total amount of capital involved in liquidations on account of mergers larger for the six-year period of 1933–38 as compared to the six years from 1927 through 1933 — 1,799,900,000 RM as compared to 1,642,700,000 RM — but also the average size of the firm undergoing mergers was considerably greater, having a capital stock of 5,210,000 RM for the years of the Nazi regime but of less than 3,740,000 RM for the pre-Nazi period. The most important amalgamations took place in oil, electric power, steel, and chemicals.[9]

Another aspect of corporate policy is seen in several of the reorganizations which took place as a result of Hitler's imperial successes in Austria and Czechoslovakia. The expansion of the Hermann Goering Works by absorbing the spoils of territorial conquest is a remarkable story and will be discussed in more detail in an adjoining section. Variations on the same theme, although in a less spectacular way, are the foreign holdings taken over by two of the largest German banks, the Deutsche Bank and the Dresdner Bank.

After the *Anschluss* of Austria and Germany, the German government obtained a majority control of the Austrian Creditanstalt Bank through the "Viag" or Vereinigte Industrie-

---

Wirtschaftliche Mitteilungen, November 30, 1940). The total amount of capital stock involved in new firms, however, declined as follows:

NEW FIRMS
(Capital stock in 1,000 RM)

1926	214,000	1931	543,400	1935	85,400
1927	350,500	1932	93,400	1936	36,200
1928	329,243	1933	298,700	1937	163,400
1929	507,000	1934	212,800	1938	82,400
1930	559,700			1939	201,000

[9] Computed from data in *Stat. Jahrb.* for various years and *Vjh. Stat. d. D. R.*, 1939, I, 115. Information regarding specific companies in amalgamations is contained in *W. und S.* (see particularly XIX, 1939, p. 237) and in *Vjh. Stat. d. D. R.*, especially 1935, I, 138, and in the *Economist* (for example, January 20, 1934, p. 117).

Unternehmungen, A.G., which is the government's holding company for its industrial and banking interests. At the same time the Austrian Industrie-Kredit, A.G., an industrial holding company controlled by the Creditanstalt, was merged with it. In January 1939 "Viag" sold 25 per cent of the nominal capital of the Creditanstalt to the Deutsche Bank.[10]

The second largest bank in Austria—the Laenderbank, which had its seat in Paris, although its branches in Austria accounted for some 80 per cent of its business — and the Vienna branch of the Czechoslovakia Živnostenská Banka were joined with the Mercurbank, a subsidiary of the Dresdner Bank.[11]

A careful and thorough survey would undoubtedly reveal other extensions of German industrial empires. These extensions, unfortunately, are not reflected in the official statistics for capital stock, bankruptcies, new firms, and mergers in Germany. In many cases the enlargement of control on the part of German corporations occurred as the result of obtaining controlling interest in other German corporations or in foreign companies. Despite the fact that a firm may be controlled by another, its identity is kept separate so long as each has a different corporate charter. This circumstance, obviously, obscures the extent of actual concentration of corporate control.

Mergers and concentration of control, an increase in the average size of firms along with a decline in new firms and bankruptcies, have significance for the functioning of the whole economy. Stability in the identity of corporations and a smaller number of larger firms assist government regulation at the same time that they are in large part a result of government action. Concentration of control and the growth in average size of corporations, on the other hand, have enhanced the bargaining power of heavy industry.

[10] League of Nations, *Money and Banking, 1938/39*, II, 78.
[11] *Ibid.*

Bankruptcies in a liberal economy fulfill the function of ridding the economy of inefficient units, while new firms play the role of increasing output where demand is great and profits above normal. Do declining bankruptcies in Germany indicate inefficiency? It is obvious that the present tendency toward stability tends to freeze the *status quo* and protect monopolistic elements. A stable price policy has prevented exorbitant war profits, but the government's control over the capital market has in part protected existing interests against new competing firms.

Nazi economists maintain that competition is wasteful, since it may involve entry into an industry because of inaccurate knowledge or ignorance regarding the real situation. Propaganda, official pressure, and business research carried on by the Estate of Industry and Trade are now used to encourage efficiency. A profit differential is the reward for increased efficiency; bankruptcy may be the punishment only in an extreme case of continued inefficiency in a branch of industry considered unimportant for Nazi world supremacy.

THE HERMANN GOERING WORKS

One of the most important mergers resulting from German victories occurred in the Hermann Goering Works. An analysis of its expansion may well give the key to the true significance of the Nazi economy. To call this enterprise a state concern adds little to an understanding of its role and significance, for it is a mixed-company form. What is more important, however, "under the rule of the Nazi elite public responsibility has been abolished and public property has become, to some extent, the hunting ground of the party bosses. The most ambitious among them, such as Hitler, Goering and Ley, have amassed industrial kingdoms in addition to their political castles, Hitler in the publishing business, Goering in iron and steel, Ley in the multicolored assortment of his two Labor Front holding companies." [12]

The Hermann Goering Works started with a capital of 5 million RM, provided by the government for the purpose of developing low-grade ore. With the help of private capital and additional public resources, its capital stock was raised to 400 million RM in 1938, making it the third ranking corporation in Germany. Backed by the government, it spread from iron-ore production into other fields. The greatest opportunity came with the conquest of Austria, where a branch of the Goering trust was established at Linz, a month after the entry of the German troops.[13] In addition, the Viag, the holding company of the Reich, which took over controlling interest in the most important industrial enterprises in Austria — by means of acquiring the majority of the Creditanstalt shares — shifted some of its shares to the Goering trust.[14] Although these shares usually represented only minority holdings in various Austrian industries, they nevertheless also represented an expansion of the power of the Goering Works, which thus acquired important interests in the following Austrian corporations:[15]

1.	Steyr-Daimler-Puch, A.G.	Leading automobile and arms manufacturer
2.	Donau Dampfschiffahrtsgesellschaft, A.G.	Danubian shipping company
3.	Machinen- und Waggonbau Fabrik, A.G.	Important railroad manufacture
4.	Bau, A.G. Negrelli	Building company
5.	Steirische Gusstahlwerke, A.G.	Special steel works
6.	Feinstahlwerke Traisen, A.G.	Special steel works
7.	Pauker Werke, A.G.	Light machine tools

[12] Kurt Lachmann, "The Hermann Goering Works," *Social Research*, February 1941, pp. 37–38.
[13] Lachmann, *loc. cit.*, p. 30.
[14] Karl Schattendorn, "Konzern Hermann Goering Werke," *Der Wirtschaftsring*, December 23, 1938.
[15] *D.V.*, October 6, 1939.

In addition, the Hermann Goering Works obtained shares in the Veitscher Magnesitwerke, A.G., and the Vitkovice works of Czechoslovakia, formerly owned in large part by the Gutmanns of Vienna and the Rothschilds of Vienna and London. Baron Louis Rothschild, who was imprisoned after the Nazis occupied Vienna, was held for more than a year until he signed away his rights to Vitkovice; the Goering trust took control later, in June 1939.[16] These various participations have been concentrated in a 100,000,000 RM holding concern by the name of A.G. Reichswerke "Hermann Goering."

The German Steel Trust, which had for many years owned a majority control of the Alpine Montangesellschaft, the largest Austrian iron-ore company, sold its majority control to the Goering Works in March 1939.[17] A subsidiary branch of the Goering Works was then amalgamated with the Alpine under the name Alpine Montan, A.G., "Hermann Goering." Lachmann considers this transaction a considerable victory for the Goering combine over the big steel men of the Ruhr.[18] The Goering combine also obtained control of the company Fanto, A.G., Vienna, and formed the Benzolvertrieb der Reichswerke "Hermann Goering," A.G., Vienna.

Military victories in Czechoslovakia brought additional trophies to the Goering trust. Half of the lignite mines of northern Czechoslovakia "were either confiscated or bought under pressure and then amalgamated into the Sudetenländische Bergbau, A.G. under the combined leadership of the Viag and the Göring trust, with a capital of 50 million marks."[19]

The Czechoslovakia Skoda works and the Brno arms factory were taken over by the Nazi government, and the only link

[16] *New York Times*, June 23, 1939.
[17] *Der Wirtschaftsring*, June 9, 1939.
[18] Lachmann, *loc. cit.*, p. 31.
[19] Lachmann, *loc. cit.*, pp. 32–33.

with the Goering Works is an interlocking directorate — Karl Rasche of the Dresdner Bank of Berlin is a director of the Skoda and Brno concerns and at the same time a director of the Goering-dominated Sudetenlaendische Bergbau, A.G.

The holdings of the Goering Works have also been extended to Rumania and Norway. Information about property changes in France and Poland is not available. Albert Goering, a nephew of Hermann, and Guido Schmidt, manager of the Austrian Goering Works, in the summer of 1940 were made members of the board of directors of the Reshitzsiron works, the largest iron and steel works of Rumania. An anti-Semitic campaign one year earlier had resulted in the ousting of the general manager and chief owner of Reshitza.[20] Albert Goering is also on the board of another Rumania armament factory, the Copsa Mica.[21]

The Goering trust has not limited its control through mergers and the acquisition of majority control in other companies to foreign countries. Part of the first steps of its domestic operations have already been described. Additional expansion was achieved by acquiring shares from the Viag. In this way, as well as by direct purchase, it took over the control of the following companies:

Middle of 1938	Rheinmetall-Borsig
End of 1939	Preussengrube, A.G.
End of 1939	Gewerkschaft Sachsen
Beginning 1940	Rheinisch-Westfaelische Industrie Beteiligungs, A.G.
	(Thyssen GmbH renamed — Fritz Thyssen's most important holdings were in the Vereinigte Stahlwerke)

The profit return from this vast Goering empire is limited, presumably by the corporation law which restricts salaries of

---

[20] *New York Times*, August 1, 1940.
[21] Lachmann, *loc. cit.*, p. 35.

corporate officials, to an amount commensurate with service and also by the Loan Stock Fund law, which limits cash dividends to 6 per cent. An extensive empire such as this would nevertheless bring unusual total returns even though the rate were limited to 6 per cent. It should, moreover, be remembered that amounts above 6 per cent may also be used for additional capital accumulation in the concern itself. After a certain point, Goering, particularly Goering, is probably not so much interested in additional consumable income as he is in the power which ownership of capital gives him. He is the archetype of a new species, half politician and half industrial overlord. Ownership secures the power acquired by political office and has strategic importance in maintaining his economic and social status if he should lose his position as second most high executioner.

V

CARTELS

THE TYPE of cartel organization which grew up in the basic industries in Germany at the turn of the century was similar in many respects to the economic structure provided by the Nazis for the whole economy. In the basic industries, though rarely elsewhere in the economy, production was distributed by quota among the member-producers, and concern with technical efficiency — and thereby increase in profits at the set cartel price — was the chief activity left to the entrepreneur.

Other types of cartel agreements prevailed in other industries. Conditions cartels regulated conditions and terms of sales; the calculations cartels were concerned merely with methods of calculating or computing costs and prices; regional cartels distributed regional markets assigned to the individual producers; production cartels attempted to control price through agreement not to expand plant or increase plant efficiency and to close down certain plants or to shorten hours of labor; submission or contract cartels determined which members should submit lowest bids for given contracts; patent cartels provided for exchange and pooling of patent and copyright holdings. There were other types, but these were the most important. Centralization of control was carried further by means of syndicates in which individual members surrendered enterprise identity for the duration of the agreement. There were continual struggles among the members themselves to enlarge quotas at the expense of fellow members. Moreover, firms which developed in size and importance outside of the cartel agreement proved to be a constant source of cutthroat competition. Cartel conflicts were in many cases brought to peace, and new agreements reëstablished under the benevolent support of government. Thus, during the first World War, controls within the war economy were founded upon cartelization. The Nazi system employed it on an even larger scale.

Ten years before the Nazis seized power, German cartel agreements had the same legal status as any other private contract. The special cartel law passed in 1923 was predicated on the assumption that cartel, price, and quota agreements among producers were not undesirable in themselves. Under this legislation, a special cartel court was established as an independent tribunal composed of professional economic court judges and laymen in connection with the Reich economic court. The Minister of Economic Affairs had the right to initiate suits in the cartel court for violations of cartel or market agreements. A firm might withdraw from a cartel agreement if there was a valid reason. A valid reason was defined as any unfair restriction of production, sale, or price-fixing, and the cartel court determined what activities were "unfair." Without the consent of the court, no cartel agreement could be broken.

The remodeling which the Act of 1923 received at the hands of the Nazis in the amendment of 1933 strengthened both the control of the Minister of Economic Affairs over cartel activities and the power of the cartel form. The amendment has transferred the vital powers of the cartel court to the Minister, who may himself determine if cartel agreements or individual decisions are unfair.[1] Under the new legislation, restrictions on entrepreneurial activity have not been considered unfair if the entrepreneur has engaged in under-cost selling or cutthroat competition. Muellensiefen, an authority on German cartels,

---

[1] *Rgbl.*, 1933, I, 487, amendment to the Cartel Act of 1923.

regards the amended definition of "unfair restriction" as a change from tolerance to positive approval of the cartel.[2]

The powers of the Minister of Economic Affairs over existing cartels have also been supplemented by his powers to compel outsiders to join a cartel in the event that "market regulation by voluntary agreements is defeated by too strong an emphasis of individual over collective interests . . ." and likewise by the power to "forbid the establishment of new enterprises and the expansion of existing establishments."[3]

The Compulsory Cartel Law enables the Minister to compel or prohibit changes in cartel agreements, prohibit the setting up or enlargement of firms, restrict the use of machinery, and determine the rights and duties of cartel members. And he can make use of his powers whenever he considers it necessary for the interests of the enterprises concerned, as well as for the benefit of the national economy.

The Minister of Economic Affairs under the new grant of powers has limited production and new plant construction for varying lengths of time in the following industries: jute weaving, paper and pulp, textile goods, cement and hollow glass, cigars and cigar boxes, clocks and watches (with the exception of wrist watches), nitrogen, superphosphates, stone materials, peat moss, radio, smoking tobacco, horseshoes, hosiery dyeing, rubber tires, white lead, red oxide of lead, white zinc, lithophone, staining and earth dyes, pressed and rolled lead products, tubing and insulation tubes.[4] From 1933 to 1935 new retail shops were established only under license, thus protecting the existing interests. In August 1933 the cigar-making industry was forbidden to introduce labor-saving machines; in June 1934 the extension of large-scale plants producing smok-

---

[2] Heinz Muellensiefen, *Vor der Kartelpolitik zur Marktordnung und Preisueberwachung* (Berlin, 1935), p. 8.
[3] *Rgbl.*, 1933, I, 488.
[4] Brady, *op. cit.*, p. 306.

ing tobacco was prohibited, and again, in 1933 and 1934, a ban was placed on the introduction of automatic machinery in chemicals and hollow glassware.[5] Until 1936, however, restrictive provisions were applied chiefly to the capital goods industries. After 1936 heavy industry, armaments, and *Ersatz* industries were given priority in the capital market.

The bans, which were placed on technical improvements during the first year or so, did not receive universal approval and there was considerable controversy over the effects of labor-saving machinery on employment. Nonetheless, the thing that proved decisive was the practical consideration of the difficulties surrounding rapid reabsorption of displaced labor when unemployment was extensive. The various embargoes on private investment retarded the expansion of industries which would otherwise have profited from new equipment orders, but at the same time the development of existing firms with large surplus capacity was favored by state orders.

The widespread supervision of cartels granted under these acts at first sight appears to mean official interference with the conduct of business, but has turned out in reality to assist business men in eliminating the "chiselers" who undermine "fair" trade and "spoil the market." Moreover, the policies of the state officials have been enforced through the machinery of the Estate of Trade and Industry and the interweaving between leading cartels in each field and the groups of the Estate has been very close so far as officers are concerned.[6] The Compulsory Cartel Act states specifically that the law should not serve as a basis for a planned economy and that governmental policy toward the cartels should attempt not only to eliminate cutthroat competition among producers but also to stabilize prices, high enough to enable industry to make reason-

[5] Guillebaud, *op. cit.*, p. 56.
[6] Heinz Muellensiefen, *Gruppenaufgaben bei der Wirtschaftlichkeitfoerderung, Marktordnung, und Kartellaufsicht* (Stuttgart, 1937), p. 89.

able profits and low enough to facilitate the government program.[7]

It is necessary for the groups to be completely independent of the cartels in order effectively to play the role of censor or regulator on behalf of the Minister of Economic Affairs. The November 1936 stop-price decree, in an attempt to get this independence, stipulated that "the director of a group in the Estate of Trade and Industry who is at the same time the leader of a cartel cannot in principle continue such personal unions." Performance as head of a governmental division and as head of a cartel was to have ended on April 1, 1937, but more than two years after the decree there had been few changes in personnel.

The cartels have from time to time faced the hostility of certain members of the Nazi party, and the role which they have played in price control has changed from time to time. But the changing importance of this role has never interfered with growth in the number of cartels. According to official estimates, the number of cartels has increased from 2,000 in 1925, to 2,200 in 1935, and 2,500 in 1936.[8] World War II actually strengthened the cartel system. The biggest victory was in the coal industry. At the outbreak of the war the German coal industry was turned over to Paul Walter, a former aide of Robert Ley, Labor Front "leader." Walter was given the title of Reich Coal Commissar and empowered to organize both coal producers and coal retailers into a state-operated syndicate. The syndicate was not a success and was dissolved. In its place, coal producers themselves formed the Reich Coal Association, modeled along cartel lines. At the head of this

[7] *Frankfurter Zeitung*, July 15, 16, 1933. The importance of this act for the regulation of economic activity is discussed by Heinz Muellensiefen, *Kartellrecht einschliesslich neuer Kartellaufsicht, Preisbildung, Schiedsgerichtsbarkeit* (Berlin, 1938).

[8] I.f.K., *Weekly Report*, Supplement, November 2, 1938.

association they placed Paul Pleiger, a business man and general manager of the Hermann Goering Works. The *Voelkischer Beobachter*, the official party paper, gave Pleiger its blessing, called the new coal organization "an extension of initiative," and proudly boasted: "The entrepreneur today, in contrast to the time of democracy, no longer looks into an insecure future." [9]

[9] *Time*, April 7, 1941, p. 79.

VI

INDUSTRIAL PRICE POLICY

THE stabilization of prices resulting from Nazi governmental intervention and continuing in the face of full employment and dynastic needs must be regarded as a remarkable achievement, unique in economic history since the industrial revolution. A series of increasingly comprehensive decrees culminated in the price-stop decree of November 1936. When the war broke out Germany had had the experience of almost three years of centralized and comprehensive price controls. This experience enabled the German economy to pursue war with only minor worries connected with prices.

Price control in Germany, however, was not an innovation of the Nazis but dates in recent times from the Emergency Decree of 1931, when prices of cartel products were lowered and a commissioner for the supervision of prices was introduced. At this time, the government set up an office for the supervision of prices, but for the most part, price regulation was enforced through local police authorities subject to the local governmental officials.

Hitler, however, abolished the office of price commissioner by an act of July 15, 1933, simultaneously with the enactment of the new cartel law. Later, however — from November 1934 to July 1935 and from November 1936 on — it was found necessary to have a separate official for this job. During 1933 and part of 1934 the cartels were the main instruments for supervision of prices, and attempted to enforce the government decree of 1933 which prohibited a rise in the prices of commodities used in public works.

INDUSTRIAL PRICE POLICY

Prohibition against price increases was extended on May 16, 1934, to all necessities.[1] Nonetheless, price indexes reflected a slow and continuous rise, and the police closed retail stores in various parts of the country as a reprisal against unauthorized price increases. By further decrees of August 1934 and December 1934 the ban on increase of prices was applied to all goods and commercial services unless controlled by the Reich Food Estate and the Chamber of Culture, or approved for inland shipping by the Minister of Transport. And in November 1934 Hitler appointed a special price commissioner. During 1934 cartels were required to notify the Price Commissioner of all price agreements and restrictions and obtain his approval for any increases. As a result of numerous "null and void" orders from the Price Commissioner, cutthroat competition appeared, and a new decree, aimed at preventing "unfair" competition, was issued in December 1934. Unfair competition was defined so as to include prices out of line with those of normally conducted business and at the same time having no relation to costs. It also included deliberate failure to meet responsibilities toward employees, creditors, or the state.

According to regulations laid down in March 1935, it was illegal for firms submitting tenders for public contracts to make agreements or uniform recommendations concerning prices without the approval of the Price Commissioner. Special price instructions were issued for textile raw materials and nonferrous metals where foreign trade transactions were involved.

The effectiveness of these decrees was greatest in the cartel sector. In other areas not subject to cartel agreements or to regulation of conditions of re-sale, as in the case of branded goods, price regulation was more difficult. The existence of machinery for the control of prices undoubtedly delayed and moderated price increases, since entrepreneurs wished to avoid special investigation and regulation. The position was suf-

[1] *Rgbl.*, 1934, I, 389.

ficiently stable so that the Price Commissioner's office was abolished in July 1935.

During 1936, however, a marked change set in as a result of the increase in prices of world raw materials. The first rise appeared in the "free" price section, and, although the increases were not great, they were widespread. Enforcement of price decrees was exceedingly difficult when everybody was tempted to increase prices, and the disadvantages of the absence of a central authority became apparent. Purchasers were ready to offer better terms in many cases if they could only secure prompt delivery; shortages had appeared. Conflicts of jurisdiction and divergences over interpretation of policy between different states created vexatious consequences. Labor was growing restive in the face of an obvious decline in real wages.

In September 1936 came the announcement of the Second Four-Year Plan with its additional strain upon resources, and along with it came centralized price control. A commissioner for price formation rather than price supervision was appointed on this occasion. He received supreme control over all prices in Germany — subject only to General Goering. The control of all price-regulating offices (including those of the Food Estate), as well as the supervision of cartel prices, was centralized under his direction.

A price-stop decree of November 1936 prohibited all increases above the level of October 17, 1936, except with the consent of the commissioner for price formation, Herr Wagner. For textiles, the price base chosen was the level on November 30, 1936. A remarkable aspect of this decree was its comprehensiveness, for it covered everything except wages and the capital and money markets. Every price from club subscriptions and insurance contributions to the hire of shooting rights or the fares of municipal trams was included, and asking or offering a higher price became a punishable offense.

The textile industries as well as other industries using foreign raw materials presented particularly difficult problems. An order of March 1937 permitted all textile firms without special permission to raise prices above the base level by the amount of any increase in raw material prices. At the same time, however, it required textile firms to lower selling prices by the amount of any decrease in the prices of materials. The general standard was set that selling price should equal only an amount sufficient to cover costs, plus a reasonable allowance for depreciation and profit.

A decree of July 1937 provided a similar arrangement for industries employing imports. Prices of commodities made wholly or partly from foreign imports might vary with the prices paid for the imported materials; and, in order to encourage the formation of stocks, the importer was allowed to re-sell at replacement cost. Moreover, the trader or manufacturer at every stage was permitted to increase prices by the same amount. An appeal was made to every entrepreneur to examine his profit margin and to decrease profits if the margin was too high.

Although subject to the supervision of the Price Commissioner, control of imported goods prices was administered by the Exchange Control Boards. The general instructions given by the Price Commissioner to the Exchange Boards throw light on price policy and were as follows:

1. Supplies of foreign goods are to be controlled and allocated so as to prevent price increases wherever possible — despite scarcities.
2. Increased import prices are to be absorbed so far as possible by reduction of trading margins.
3. Profit rates should represent only reasonable returns and must not rise with an increase in prices.

The price-stop decree of 1936, which was further extended in September 1939, contained three elements of flexibility. Prices which had been decreased after October 17, 1936, might be raised again without permission if they were not increased above the level of October 17; the commissioner had power to lower prices by decree; price rises could be granted as exceptions if a firm's existence was endangered without it — providing that the firm was not obviously inefficient or that the high costs were not attributable to high wages.

The Price Office was swamped with requests for exceptions from the price-stop decree, and the Price Commissioner in reviewing his work at the end of his year in office stated that he gave exceptions freely wherever he considered them proper. According to C. W. Guillebaud, Herr Wagner had been chosen as Price Commissioner for his ability to say no, as well as for his sound common sense. In any case, as a result of Wagner's administration, wholesale prices of consumption goods rose only 5 per cent between October 1936 and October 1937 and by approximately 3 per cent between April 1937 and April 1938. A large part of the rise in the first year occurred as the result of a rise in imported materials or the utilization of high-cost domestic substitutes.

Price reductions were not large and were very much limited in sphere, but determination of these reductions illustrates the manner in which control worked. Reductions from 5 to 10 per cent were made in the retail prices of certain branded goods, chiefly electrical equipment, chemical goods, and clocks and watches. Prices of artificial fertilizers were also reduced as much as 25 per cent; staple fibre by 9 per cent; aluminum by 8 per cent; flat glass by 10 per cent.

According to the German Business Research Institute, the saving to consumers' income from these measures was estimated to be only 300 million RM out of a yearly retail volume of approximately 31 billion RM. Their psychological effect on consumers was to encourage the use of branded commodities, which in turn facilitated control during war times.

Manufacturers, wholesalers, and retailers shared the burden

of the price cuts, but not in equal proportions. Whereas in 1931 price reductions were imposed upon industries and traders from above, uniformly and arbitrarily, the price cuts of 1937 were the outcome of negotiations between the Price Commissioner and various units in the Estate. Moreover, reductions occurred in areas where prices had fallen relatively little during the slump. Many party members felt that price cuts were entirely too modest.

Price control which was watertight on paper showed certain loopholes in practice. Manufacturers might obtain the higher prices permitted for sales in small quantities by refusing to sell in larger quantities, by requiring consumers to purchase an unwanted commodity in order to obtain an article which was needed, and by tightening the terms of credit. It is extremely difficult to prevent secret alteration of quality. Thus evasion, as well as legal provisions for flexibility, has somewhat eased the rigorous control over the manufacturer.

The Second Four-Year Plan which brought ascendancy of power to Goering also created some confusion in the relation between the Estate of Industry and Trade and the cartels in regard to prices. Previous to this decree, the Estate activities were excluded from the realm of price determination. Dr. Schacht, who viewed with some dismay the increased power of Goering, set forth a decree on November 12, 1936, which might have increased the importance of the Estate. Schacht apparently hoped in this way to counterbalance Goering's power. The groups and the chambers were charged with seeing that cartel activity accorded with the economic policies of the national state. Moreover, they were given the opportunity of filing complaints against cartel agreements with the Minister of Economic Affairs. Schacht maintained that autarchy brought diminishing returns at an early stage for industry and that cartels further restricted elasticity by attempting to maintain the *status quo* at a time when radical technical changes were needed. He insisted that the new four-year plan required experimentation and innovation which the cartel system restricted. For these reasons he decreed that the Estate of Industry and Trade should be empowered to assist in price fixing.

Firms through the Estate were urged to keep profit and loss accounts with a view to avoiding unnecessary price increases. Particularly in the handicrafts group, publicity has favored proper bookkeeping methods. The Estate also helps to instill a spirit of business honor in regard to quality and does propaganda work on behalf of clarity in prices and description of commodities. Through assistance of one of the economic divisions in the Estate each enterprise may be induced to form the type of organization looked upon as most efficient by its particular economic division in the Estate. In this way rationalization and decreased prices have been encouraged.

Exempt from regulation or any kind of control or influence by the Estate were international cartels as well as producers of trade-marked and branded articles.

The devices for stabilizing prices which the government administers by means of the Estate of Trade and Industry, the cartels, and the Price Commissioner's office, are summarized in the following outline:

1. Control of supply and costs
    a. Quotas and priorities on raw materials and imports
    b. Rationalization
    c. Attempts to establish fixed profit margins
    d. Regulation of wage rates and transportation costs
    e. Efforts to establish price of average firm for the whole of an industry
    f. Publicity and education in regard to costs and bookkeeping
2. Control of demand
    a. Propaganda for use of German-made articles which are easily produced
    b. Rationing
    c. Attempts to maintain quality through standardization and trade-marking

INDUSTRIAL PRICE POLICY

The outstanding features of the German price control mechanism have been not only its many-sided characteristics, as presented in the outline above, but also its lack of any unified application of policies. Economic literature in German periodicals is fairly unanimous in praising price policy for being realistic rather than for attempting to achieve theoretical goals set up by equilibrium theory.[2] German economists also appear to agree that:

1. Free competition and price as a principle of organizing the economy is finished.
2. The principle of "performance competition" should be set up as a basis of price regulation.
3. Efforts to control prices and economic life should be made as flexible as possible.
4. Changes in fundamental economic organization brought about by the Compulsory Cartel Act and the Amendment to the Cartel Act of 1923 are particularly praiseworthy.[3]

With the outbreak of war, the 1936 price-stop decree was extended. The War Emergency Act of September 3, 1939, decreed that all price calculations should be reëxamined and that allowance for war risks was not to be included in price, but that additions for increased costs might be made.[4] Under no circumstances were consumers' goods prices to be lowered, and maximum profit rates were to be established in the textile retail trade and for other consumer articles.[5] The principle of uniform prices for an industry was established on behalf of industrial efficiency, the uniform price to equal average costs (including normal profits) of all firms in the industry (excluding "hothouse" — government subsidized firms).[6] Government

[2] Carl Billich, "Vier Jahre nationalsozialistische Kartellpolitik," *D.V.*, 1937, p. 2527.
[3] Elisabeth Linhart, "Wettbewerbstheorien-Wettbewerbspolitik," I.f.K., *Vierteljahrshefte zur Wirtschaftsforschung*, 1939/40, Heft 1, pp. 138–150.
[4] I.f.K., *Weekly Report*, March 13, 1940.
[5] *D.V.*, September 29, 1939.
[6] *Der Vierjahresplan*, September 29, 1939.

subsidies were extended to firms producing important commodities under high cost conditions. Certain war taxes, such as the tax on beer, may be included in the price but must be indicated in the price schedule. A novel provision has been made for the control of soap prices: 90 per cent of differential rent must be pooled in a central fund which is to be used to subsidize the costs of raw materials; the remaining 10 per cent may be retained by a firm as an efficiency bonus.[7] By the time of the war 150,000 industrial establishments were being regularly circularized by the Statistical Office in the interests of war planning.

The organs of the Estate, termed by the Nazis organs of economic self-government, gave rise to much and almost violent discussion as militarization of the economy was established. The Estate agitated for the right to be entrusted with suballocation of raw materials. This was refused as a general principle, but the "self-governing" units were permitted control over certain materials: iron for repairs, driving-belts, and lubricating oils. Allocations of materials ceased to be granted for a specified time and were established on the principle that "raw materials follow the orders." This gave greater flexibility to priorities.

Certain firms were designated as "war plants" under direct army control. This arrangement was known as the "w" system. The object was to choose the most efficient firms, which were to be specially favored by priorities, and to shut down the others. In overcrowded industries inefficient firms are encouraged to liquidate, and efforts are being made to finds ways of fitting non-"w" firms into the war economy. Army orders are placed directly by the army and air ministries, which also directly assist the financing of "w" plants.[8]

Following the outbreak of war the role of trade-marks in

[7] *D.V.*, October 13, 1939.
[8] H. W. Singer, "The German War Economy in the Light of German Economic Periodicals," *Economic Journal*, December 1940, pp. 536–537.

price controls created an interesting controversy. Trade-marks were abolished in the following industries: soap, washing materials, biscuits, cornflakes, and shoes. *Der deutsche Volkswirt* protested that this was going too far and suggested that, in the case of shoes, trade-marks were a stimulus to efficient and honest production.[9] In November 1940 the same periodical ran an entire issue on the virtues of trade-marks and brands. Trade-marks were viewed as of particular value in enforcing standardization of quality, and hence facilitating price regulation. A decree of October 5, 1940, prohibited the "watering-down" of trade-marked articles; retailers and sellers were to be punished for handling such goods. According to the same decree, advertising costs might be covered in prices, but such expenditures should not be "excessive." Advertising must not to be used merely to widen the market for an old product or to introduce new products where existing commodities were adequate. It was also proposed that consumers' goods, particularly hard hit as a result of the war, should set up an advertising reserve for large expenditures after the war.[10]

A wage-stop decree buttressed the price-stop decree, and general propaganda pointed the finger of shame at what was called "labor-war-profiteering." In addition to the wage-stop decree, wage rates were brought down by a variety of other measures. The amount of work performed for piece rates was extended; seniority privileges, promises of bonuses, and profit-sharing schemes were employed rather than higher rates.[11] Wages in nonessential industries were actually lowered below the rate of the depression. Finer classifications of skills and corresponding wage rates were set up.[12] At first, overtime pay was abolished, but this proved so unsuccessful that later it was

[9] Singer, *loc. cit.*
[10] *D.V.*, November 15, 1940.
[11] *D.V.*, October 27, 1939.
[12] *Der Vierjahresplan*, September 15, 1939.

reëstablished. Overtime is now paid to the worker after the eleventh hour; overtime for the ninth and tenth hours each day is paid by the employer to the government.[13] Regimentation of labor supply, discussed in the chapter on labor, also assisted the stable wage policy.

Another aspect of price controls was the Spartan regime imposed on German consumers. Among the industrial products rationed from August 27, 1939, were soap, shoes, and nearly all textile products. On November 14, 1939, a clothing card was introduced by the government. The card assigned a certain number of points for every article of clothing from neckties to underwear, and each person was allowed to buy clothing up to 100 points per year. A suit counted as 60 points, shorts 15, pajamas 25, and a shirt 20.[14] Overcoats were made available only if a worn overcoat was surrendered in exchange. Although other industrial goods were not rationed, many of them gradually disappeared from store shelves, and new supplies could not be secured.

It is the boast of the Nazis, however, that rationing in the present war as contrasted with the first World War is conducted fairly and in an organized manner. Luxuries were naturally the first commodities hit by the stringency of the war, and a new classification of war luxuries — commodities which are necessities in peace but luxuries in war — has made its appearance.[15] Motor cars may be used only for public purposes, and even then only small cars are to be driven. Misuse of the "public purpose" privilege is subject to heavy fine and is interpreted to cover traveling in an automobile where railway service is adequate, driving through town traffic, and taking joy rides.[16] A system is being established whereby

[13] I.f.K., *Weekly Report*, March 13, 1940.
[14] *Voelkischer Beobachter*, November 15, 1939.
[15] *D.V.*, September 8, 1939.
[16] *D.V.*, September 29, 1939.

INDUSTRIAL PRICE POLICY

buyers must go to certain specified shopkeepers for commodities. This enables the government to keep track of hoarding, and eliminates the necessity for large numbers of shopkeepers, who thus may enter the ranks of labor.

The use of metals except on war commodities has been entirely stopped.[17] The use of substitute materials is encouraged by a set of "Special Advice Bureaux." Differential railroad rates, lower than average, are set for high-cost iron-ore fields in Bavaria. Construction by the Reich on motor roads has been cut 50 per cent, while building construction has been reduced 40 per cent. General construction by local [18] and state governments has undergone a cut of 50 per cent.

Increased war activity enlarged the difficulties of price control, but price indexes present a picture of controllable and controlled development. The two greatest problems arose from increased utilization of plants and increased price of foreign imports and domestic substitutes.

[17] *D.V.*, September 8, 1939.
[18] Singer, *loc. cit.*, p. 543.

VII

CONTROL OF FOREIGN EXCHANGE AND FOREIGN TRADE

EXCHANGE CONTROL before the advent of the Nazis had been administered largely to limit withdrawals of foreign credits; after their rise to power, it developed into the fulcrum of the central control of all important primary commodities. From this central citadel, the government controlled the source, quantity, and use of raw materials and hence controlled business itself. But government interference was restrictive only for industries outside armament and *Ersatz* productions. Armament and *Ersatz* firms held priorities over all others during pre-war and war years, while after the outbreak of hostilities, Nazi plans for the reorganization of Europe included measures for closing down competing industrial plants in defeated countries.

If exchange control had been aimed merely at recovery in domestic industry, it might easily have been relaxed by 1935 and possibly before.[1] It was thus largely in preparation for war that controls were extended, strengthened, and centralized.

[1] According to Howard Ellis, "exchange control from the monetary and financial angles was superfluous as early as 1933 in all probability, but by 1935 for a certainty." He bases his case for devaluation as early as 1933 on the following circumstances: "A great deal of German trade already proceeded over *de facto* devalued rates, the capital flight was under control, debt services had proceeded satisfactorily, there were clear signs of economic recovery in Western Europe." Reparations had largely been annulled at Lausanne in 1932, and the foreign debt service was reduced by the Standstill agreement. The uncertain extent of American devaluation was the only large obstacle. ("Exchange Control in Germany," *Quarterly Journal of Economics*, supplement, vol. LIV, 1940, pt. II, pp. 126–127.)

In mid-1935 Austria, with a per capita foreign indebtedness as heavy as

CONTROL OF FOREIGN EXCHANGE

Pressure from protected industries dovetailed nicely with the interests of bureaucrats and party members. We may now proceed to consider the nature of these controls in their development from a limited sphere into a Leviathan of regulations.

Administration of exchange control and the mass of legal provisions and decisions under which it operates is not a main concern of economists. But a brief description of these provisions is indispensable to an understanding of Nazi institutions. Since the introduction of exchange control in the crisis of July 1931, there have been three general exchange control laws, upwards of fifty separate decrees of amendments, and more than five hundred administrative rulings, not to mention clearing, compensation, and payment agreements with partner countries. A general recodification of all exchange regulations finally was decreed on January 1, 1939, for Greater Germany.

Foreign exchange control may be understood to include all measures aimed at stabilization of the foreign exchange market. These measures in turn fall into four classifications:

1. Elimination of speculative activity and sharp variations in rates
2. Prevention of flight of capital
3. Government determination of all foreign exchange prices
4. Weapon of commercial policy, particularly of autarchic policy

---

Germany's, greater unemployment, and larger territorial losses, liberated exchange payments except for capital flight, thus taking a pioneering step towards devaluation in central Europe. Germany, with a key position in the economy of central Europe, refused to follow suit.

"The gold *bloc* devaluation of 1936 offered another golden opportunity, for by that time the German economy was thoroughly controlled within, and the foreign debts reduced to half their magnitude in 1931. Since Germany did not move, it was difficult (though not impossible) for Hungary to institute a devaluation. The funding of the Hungarian debt and the resumption of effective devisen transfer in the summer of 1937 supplied another and, as it proved, a final opportunity for abolishing the fictitious gold pengo." (Ellis, *loc. cit.*, p. 183.)

Germany's course of action thus sealed the fate of the gold-*bloc* countries and necessitated the continuation of innumerable controls to maintain motley and numerous but concealed devaluations.

EMERGENCY DEVICES TO CHECK THE FLIGHT OF CAPITAL, JULY–NOVEMBER, 1931

In 1934 Hitler introduced the most rigid form of controls, but as early as 1931 measures had been taken to prevent the flight of capital. Various decrees during 1931 gave to the Reichsbank a monopoly of dealings in foreign exchange, prohibited all deviations from the official rate of exchange, and abolished forward transactions in exchange.[2] Owners of foreign exchange were required to sell their exchange to the central bank. Foreign securities could still be purchased but had to be registered. The decrees also permitted purchase of exchange with German securities or devisen. The postal authorities prohibited sending of money through the mails, but nothing was said explicitly about a general embargo on the exportation of German money. Finally, German residents with ownership in a foreign enterprise were required to register their holdings, provided their equity represented 50 per cent of the total amount and there were no more than five owners in the enterprise.

The first formal Devisen law of August 1, 1931, attempted to plug the gaps in the earlier decrees of that year. Accordingly, purchase of *Devisen* was possible only at authorized banks and upon presentation of a certificate issued by the newly erected Devisen offices. A similar certificate was a prerequisite for purchase of foreign securities not regularly quoted on the German bourses. Permission of the Devisen office was also required for the following transactions:

1. Opening of new credits to foreigners
2. Disposal of mark accounts in Germany owned by foreigners and originating before August 4, 1931
3. Transference of marks to accounts held with firms abroad

[2] *Rgbl.*, 1931, I, 365, 366–368, 369, 373–376.

And, at last, a general embargo upon exporting domestic means of payment was introduced.

Even these measures, however, had loopholes. All prohibitions of August 1931 applied only to transactions over 3,000 RM, and control was notoriously ineffective. The continuance of smuggling resulted in the tightening of existing decrees. The amount exempted from exchange control was lowered from 3,000 to 1,000 and subsequently to 200 RM.[3] On November 2, 1931, the purchase and sale of all foreign securities was subject to the permission of the Devisen office.[4]

In the midst of the July crisis, the London conference investigated the German short-term debt situation and on September 19 finally reached a "settlement" known as the First Standstill Agreement. The basic idea of this agreement was to maintain existing credits without alteration of their original terms, but to give relief by postponing the date of payment. Stupidly, the moratorium was set at six months, although no one could reasonably expect a significant change in German conditions in such a short time. Moreover, the agreement covered only half of the total short-term debts, while call-money loans, short-term advances against securities and mortgages, and credits to agriculture, as well as the short-term debts contracted by states, municipalities, and public bodies, were excluded.

It is thus doubtful whether the standstill agreements formed a basis for the discontinuation of exchange controls. In addition, Great Britain's departure from the gold standard a few days following September 19 created additional problems. Moreover, the effect which exchange control had on foreign prices of German bonds encouraged further control. The transfer moratorium created artificially high values of German bonds at home and low values on the same securities abroad.

[3] *Rgbl.*, 1931, I, 463, 534.
[4] *Rgbl.*, 1931, I, 673–674.

This combination would have encouraged capital flight, if other causes had been missing, for repatriation offered windfall profits on exports of capital. From this situation, it followed that the Reichsbank found conversion of foreign short-term credits into long-term obligations practically impossible, but exchange control could not be withdrawn until conversion had taken place.

EXCHANGE RATES UNDER EXCHANGE CONTROL 1932 AND 1933

The pressure for capital flight and overvaluation of the mark relative to sterling necessitated rationing of foreign exchange because demand exceeded supply at the official rate. As introduced in November 1931 the rationing scheme permitted each importer to have a certain percentage of his actual exchange requirements in a base period, June 1, 1930, to July 1, 1931. At first, the importer was permitted to buy 75 per cent of the exchange purchased in the base period, but by May 1934 this had fallen to 50 per cent, and before the end of June 1934 to a day-to-day allotment based on the Reichsbank's intake. Each individual case had to be investigated; the administrative machinery was hopelessly swamped. Larger demands received attention at the expense of smaller requests for which there was no time. By the time an importer was permitted to buy exchange, he no longer wanted it; and when he asked to use it on another shipment he was refused and had to start the whole process from the beginning.

Rationing may thus be regarded as a protectionist device, for it prevented the appearance of new import firms, practically made impossible any change in relative outputs of the various firms, and thus provided a sheltered position for existing importers. Despite the protectionist aspects of this early rationing scheme, trade policy was still not the dominating theme of exchange control. The Association of German Industry, however, carried on active propaganda in favor of using regulations for this purpose.

In spite of favorable signs during 1932 — the renewal of standstill agreements with more elastic terms, the virtual abandonment of reparations at Lausanne, and the reorganization of the banks which had closed in the summer of 1931 — Germany rapidly lost her export capacity in competition with the sterling *bloc*. Hitler's adoption of a definite expansionist program of government spending further worsened the position of German prices and costs relative to foreign countries. In this situation only two alternatives existed: either "open devaluation or devaluations concealed beneath exchange control; and in the latter event, exchange control would become a permanent feature of the economy." [5]

The dwindling gold reserves and the shrinking favorable trade balances may have been engineered by the government as a pretext for reducing the foreign debt. For, as a result of the foreign trade emergency, a decree of June 9, 1933, proclaimed a transfer moratorium for long- and medium-term debts.

Under this moratorium, a scrip system was set up. German borrowers were required to pay obligations into a conversion office instead of to the foreign creditor. Out of the funds going into the conversion office, one-half was transferred to the foreign creditor in free marks (convertible into gold or foreign exchange) and the other half in scrip. Scrip could be transferred into free marks only at the Gold Discount Bank and only at a discount of 50 per cent — hence the foreigner received but 75 per cent of the amount owed him. Later, December 1933, the proportions were changed to 30 per cent free marks and 70 per cent scrip, with the scrip redeemable at 67 per cent of its full value. Interest and amortization on the Dawes loan and interest on the Young loan were to be trans-

[5] Ellis, *loc. cit.*, p. 32.

ferred in full. Moreover, various countries, notably Holland and Switzerland, forced through special arrangements under which their nationals received interest in full in return for extra purchases of German goods.

The most important aspect of this ingenious scrip system was the fact that it served as a subsidy to exporters. The Gold Discount Bank put scrip at the disposal of the exporters at a discount of 45 per cent. The exporter actually realized about 10 to 20 per cent of his total exports, the amount being computed according to the ratio:

$$\frac{\text{loss from export on account of overvaluation of mark} \times 100}{\text{discount on scrip}}.$$

But, to obtain the scrip discount, the exporter had to prove that the export was additional. Hence, for the first time, exchange control was used directly to regulate foreign trade; other measures had affected trade only indirectly by control of payment.

The moratorium, moreover, was made even more stringent on July 1, 1934, when the Reich told German creditors that they could have their choice of 40 per cent in foreign exchange as full payment of interest or amortization charges, or bonds maturing at the end of 1944, but redeemable at any time at 40 per cent of their face value. The second alternative was actually withdrawn at the beginning of November 1934. Foreign creditors could take 40 per cent or nothing. The German decree of July 1934 also announced that suspension of transfers for all German foreign medium- or long-term obligations would include the Dawes and Young loans.

However, Germany negotiated special arrangements providing for partial or complete continuation of interest payments on the debts of certain countries.[6] This was true for creditor

---

[6] The agreements with Britain, France, the Netherlands, Belgium-Luxemburg, Switzerland, Sweden, and Italy covered interest on the Dawes and Young loans; those with Britain, the Netherlands, Switzerland, Sweden,

countries in western Europe who threatened to hold back the proceeds from the sale of German goods and thus enforce payment of the debt claims of their nationals. The United States, owing to its favorable trade balance with Germany, was not in a position to exert pressure.

Clearing agreements or bilateral trade agreements had originated two years earlier as a means of overcoming the exchange difficulties of various countries in the Balkan area and in South America. In the Balkan case, Germany herself was a creditor. When these countries were unable to pay for imports in foreign exchange, the Reichsbank established special accounts for the central bank of the particular country involved. For any given special account, the amounts which German importers paid for goods brought into Germany were added, whereas the expenditures by German exporters were subtracted. The foreign debtor discharged his obligation at his central bank in domestic currency; the central bank in turn was to settle the special account at the Reichsbank.

When foreign exchange became extremely scarce early in 1934, German importers realized that they could buy goods by way of these clearing accounts without needing foreign exchange and thus avoid the rationing provisions. The importer could buy goods from any country with whom a clearing agreement had been made. Foreign suppliers already accustomed to this type of transaction happily entered into contracts on this basis. Raw materials embargoed under the rationing control could be imported through these accounts or could be brought in as semi-manufactured products. Imports from countries which had no agreements were sent through countries which did have them. And before long the credits in favor of Germany were replaced by large debts.

During 1932–34 numerous variations of the bilateral trade

---

and Belgium-Luxemburg covered other debt payments as well. (Department of Overseas Trade, *op. cit.*, pp. 21–26.)

agreements appeared, but the basic principle remained the same: German imports were paid for only by the revenue from German exports to the same country and vice versa. With a few countries, notably Great Britain, it was possible to conclude payment agreements which were more favorable than the clearing agreements, both because they were more elastic and involved less bureaucratic control of the individual trader in each country and because frequently there was a larger proportion of free exchange.

The disparity of German and foreign prices continued during the second half-year of 1933 and the first half-year of 1934. The quotas permitted the importer were rapidly cut after March 1934, and by June the Reichsbank limited quotas to its daily *Devisen* receipts. The result of this was to create considerable uncertainty as to when payment would actually be forthcoming. Quotas had thus lost all power as a method of regulating the foreign exchanges, and the last remnant of unregulated trade had disappeared.

In 1934 the government faced a situation in which the former export surplus had already been changed to an import surplus with attendant shortage of the exchange which might be used for acquiring abroad the additional imports of commodity types essential to public works and rearmament. The government extricated itself from this dilemma by diverting purchasing power from consumers' goods into producers' goods. In addition to limiting the imports of consumers' goods, profits on the sale of such goods were also limited. A decree of April 1934 forbade higher prices for textile and leather raw materials. Agreements in regard to minimum prices, profit margins, and rebates were prohibited on all products of daily need.[7]

To enforce these measures aimed at limiting consumption, the government placed textile raw material imports under con-

[7] Reichs-Kredit-Gesellschaft, *Germany's Economic Situation at the Turn of 1934/1935*, p. 36.

CONTROL OF FOREIGN EXCHANGE

trol boards, March 22, 1934.[8] Later, it established similar boards for nonferrous metals, cowhides, and rubber. The exchange control authorities also tightened the mark allowances of travelers and *émigrés*.

EXCHANGE CONTROL AS A TOTALITARIAN INSTITUTION

After an unsatisfactory period of hand-to-mouth administration, Schacht announced his "New Plan," September 1934. The New Plan abolished the system of foreign exchange allotment and substituted for it a system whereby every import transaction was subject to the approval of the proper supervisory office. Its objectives were three: (1) to hold imports strictly to the available amount of foreign exchange, (2) to confine purchases to countries which bought equivalent amounts from Germany, and (3) to give preference to certain imports of raw materials, particularly those necessary for rearmament. The New Plan, in short, was one of the principal means of making war preparations economically possible.

Subsequent development proved that the New Plan successfully fulfilled its stated aims. No basic changes in control were, therefore, necessary. Even the outbreak of war required only minor adjustments.

Twenty-seven control boards were set up to supervise the value and sources of imports, and in many cases give definite instructions as to the disposal of materials available within the home market. These boards were centralized under one Reich bureau which was staffed by representatives of various government departments and industries under the chairmanship of the president of the Reichsbank.[9] Control boards issued four kinds of permits, according as the import involved: (1) cash payments, (2) acceptance credits, (3) normal commercial credits, (4) payment and clearing agreement payments.[10]

[8] *Ibid.*
[9] Department of Overseas Trade, *op. cit.*, p. 6. [10] *Ibid.*

Exchange control thus included all payment possibilities, with the one exception of Aski marks, described in the following pages.

The operation of the control board is as follows: control boards consider the prices of the proposed imports together with other selling terms before issuing a permit. When payment of an account becomes due, importers are granted foreign exchange by the Reichsbank up to the amount of the foreign exchange certificate or permit, on proving by means of a customs clearing certificate that the goods have actually been imported. A check is then sent to the foreign exporters in the usual manner.

Large numbers of officials are employed to administer the scheme, and much extra work and expense falls upon importing firms. The centralization of control boards, mainly in Berlin, requires that many outside firms must send representatives on frequent journeys to the German capital to expedite applications by personal interviews. The operation of exchange control works to the disadvantage of small firms little known abroad. Large firms have the additional advantage of exerting greater influence on the control boards, as well as having wider opportunities for representation in the central bureau. Moreover, they have available the necessary credit for time-consuming negotiations.

Control boards were used as a weapon in the bargaining process for favorable bilateral agreements and barter deals. As far as was possible, imports were shifted to countries with which Germany had clearing agreements. The reason for this was that such agreements made it possible for German importers to pay for foreign goods in Reichsmarks rather than in foreign exchange. The other side of this policy was to restrict purchases from each non-clearing country to the amounts of German goods exported to that country. Several nations — for example, the Netherlands, France, and Switzerland — retaliated by restricting purchases from Germany. But for countries which were dependent on the German market there was little choice. They increased their imports from Germany so that their mark balances could be got out of Germany, so that German purchases might continue. This was particularly true in southeastern Europe.

A good many barter compensation deals were carried out in which ships or armaments were exchanged against surpluses such as coffee or cotton. The working of these agreements frequently meant that Germany would buy less of United States cotton and increase her purchases of South American cotton. Considerable initial dislocations in production were created in many cases because of the technical differences in raw materials.

An extensive system of private compensation agreements also developed. These agreements took many different forms; only three of the most frequently used types will be described. One form worked as follows: A German importer buying goods from the United States finds a German exporter who is unable to sell German goods to the United States because his price, at the nominal rate of exchange, is above the world price. The German importer could obtain the foreign exchange which he needed by paying the German exporter an amount sufficient to enable the exporter to accept the world price for his goods. The importer would then recoup himself for the premium paid to obtain the needed foreign exchange by increasing the selling price of the imported goods within Germany.

Another form of compensation developed along the following lines: German importer, A, offers a premium above the world price to American exporter, X, who is put in touch with another American, Y, buying goods exported by another German, B. The whole transaction would be simultaneously concluded by Y's settling with X in dollars and A with B in Reichsmarks.

The system of Aski marks which came to be used widely

represents a more generalized form of barter compensation requiring considerably less individual negotiation and pairing off of supply and demand. Under this system, a German importer paid for his goods by a credit in Aski marks. The foreign exporter, who would take into account the fact that the Aski marks could be sold only at a discount, would raise his price in marks correspondingly. He would sell them with this discount to a countryman purchasing goods from Germany at the German price, which was above the world base.

At first considerable latitude was given to Aski purchases because they did not imply the necessity for Germany to furnish foreign exchange. They were thus not subjected to the limitations imposed by the New Plan. But the huge accumulations which appeared in these arrangements became a serious menace. Luxuries or semi-luxuries were often imported in this way. Frequently, barter transactions included German goods which could just as well have been used for foreign exchange and without subsidy.

Subsequently, permission from a control board was required of all imports under Aski. A large number of regulatory decrees finally brought about the decline of compensation agreements. The first restriction, issued December 1935, excluded from Aski transaction a long list of goods in which German possessed a monopoly. A special and more stringent list was made for countries with no clearing agreements. Except for Latin America, exchange for nonessential commodities was put on a 1 to 3 ratio in favor of German exports. On February 24, 1937, sweeping restrictions decreed by the Foreign Exchange Control Bureau sounded the death knell of Aski and barter trading.[11]

As foreign exchange sources became extremely scarce in 1934, the amounts available for subsidies to export by way of the scrip discount also shrank, and other measures had to be

[11] *The Banker Magazine* (London), February 1937, p. 161.

found. A levy was therefore made on all industrial enterprises. The fund was centralized and administered by the Gold Discount Bank. With rising costs, many exports required subsidy, and the Discount Office, in determining who got the subsidy, could determine the character and volume of exports. The levy was supposedly voluntary, although in practice it could hardly be evaded.

The general levy of 1934 was merely an extension of a system which already existed in certain important industries — for example, coal, iron and steel, glass, cement, and motor vehicles. The amount which each industry pays into the general funds is determined by the Estate of Trade and Industry which collects the levy. Individual amounts are kept in strictest secrecy, but consensus of opinion outside of Germany believes it to be about one million marks annually, assessed at the rate of 2 to 5 per cent of the annual industrial turnover of any given firm.[12]

One of the most important groups of German clearings has been with southeastern Europe. By means of these clearings Germany extended her sphere of economic and political influence. The grave economic plight of the Balkan countries also helped to thrust them into the German bilateral trading system, for Germany concentrated on trade with weaker countries, which could easily be made dependent on the continuation of their German trade. An additional factor of great importance was that trade in the Balkan area was likely to remain accessible in time of war.[13]

A smaller country, once caught in the clearing network, could hardly escape without help from a large nation. During the first stage, Germany bought the available surpluses and sold in return imports which were actually needed by the small

[12] *The Banker Magazine* (London), February 1937, p. 161.
[13] Fritz Lehmann and Hans Staudinger, "Germany's Economic Mobilization for War," *The Conference Board Economic Record*, June 24, 1940.

country. Later, however, the procedure was exactly reversed: the small country had to take what Germany could spare and could sell only those things which Germany wanted. The last stage came when the small country was practically forced to adjust its economic system to German requirements. Moreover, Germany would sell desirable machinery only if she was given a voice in the management of the small country's industrial enterprises. These various stages were perhaps illustrated most clearly in the German-Rumanian agreement of March 1939.

TABLE 8
Distribution of German Imports
(*Per cent of total imports*)

	1929	1938
Western Europe, including Great Britain	27.3	19.3
United States	13.4	7.5
British Empire, excluding Great Britain	10.7	6.5
USSR and Poland	5.7	2.7
Balkan and Mediterranean countries	10.9	20.2
Scandinavian and Baltic countries	8.5	12.9
Latin America	9.4	12.6
Eastern Asia	3.1	5.1

Source: Computed from data in *W. und S.*

The resulting shift in Germany's foreign trade is illustrated in Table 8.

Although exchange control has not always achieved the goals immediately set for it, looking at the four-year period from 1934 to 1938, the system as a whole must be judged successful in the light of German objectives. Industrial production was rapidly expanded beyond the pre-depression level, even though imports and exports were not correspondingly increased, and sufficient raw materials were procured to enable the equipment of large mechanized forces. For the five year period, 1928-32, foreign purchases amounted to 18.3 per cent of German national income, whereas for 1934-38 they came to only 7.5 per cent.

At the same time, regulation made it possible to favor imports which were essential for military purposes. The effect of this war policy is seen in the disparity of decline of various groups of imports. As is clearly shown in Tables 9 and 10, all imports declined, but those needed for consumers' goods industries were sharply reduced.

TABLE 9

DECLINE IN IMPORT VOLUME

(*1934–38 as compared with 1926–30*)

	Per cent
Total imports	22
Foodstuffs	27
Finished industrial products	54
Semi-finished products	17
Raw materials	3

SOURCE: Computed from data in *Stat. Jahrb.* for various years and *W. und S.*

The war objective of international trade under German exchange control is thus seen beyond doubt in its volume, direction, and composition. Reëmployment policies necessitate

TABLE 10

IMPORT VOLUME: DECLINE IN SELECTED GROUPS OF COMMODITIES

(*Average annual volume for 1934–38 as compared with 1928–30*)

	Per cent		Per cent
Wheat	68	Copper	unchanged
Butter	36	Iron ore	16
Wool	25	Rubber	86
Cotton	24	Mineral oils	87

SOURCE: Computed from data in *Stat. Jahrb.* for various years and *W. und S.*

no exception to the statement that exchange control can be explained only in these terms. The success which the control boards had in adapting foreign trade to military activity was paid for by higher costs — the principle of comparative advantage was rejected when bilateralism became the chief guiding instrument.[14] The greatest pecuniary gain was the

[14] For an excellent analysis of this point see Ellis, *loc. cit.*, pp. 119–125.

substantial reduction in the foreign debt which Germany was able to force her creditors to accept. From mid-1931 to early 1938, indebtedness fell from 24 billion to 10 billion RM. Of this reduction, repurchases and redemptions made up 8 billions and depreciation of foreign debts through foreign devaluations represented 6 billions.[15]

[15] Ellis, *loc. cit.*, pp. 130–131.

VIII

FINANCIAL POLICIES:
CAPITAL AND MONEY MARKETS,
BANKING AND GOVERNMENT FINANCE

OUTSTANDING features of Nazi financial policies were the emphasis on early consolidation of the public debt and the observance of ordinary regulatory methods. Nationalization of the banking system was rejected. Sterilization of excess balances by open-market operations, retention of the stock exchanges, refusal to compel by a general decree a reduction in the long-term rate of interest, as well as other policies, safeguarded what remained of the individual system based on private property. The government might have utilized fully its coercive power, but it preferred to retain ordinary methods of control so long as these were compatible with maintaining the power of the Nazis at home and abroad.

On the other hand, government controls were more extensive than ever before. They were, moreover, amply sufficient to prevent runaway inflation, provided only that state loan expenditure at full employment was kept within the capacity of the productive facilities. In addition, the soundness of the German system from a monetary point of view did not rest entirely upon tightness of control over capital and investment markets, for it was also guaranteed by control over wage rates and prices.

Both the strength and the efficiency of control devices were admirably demonstrated by the lack of panic in the capital and money market at the outbreak of war. Germany had had a half dozen years of practice in a war economy. The new decrees promulgated as a result of actual war involved no fundamental recasting of credit and banking operations. Even changes in the tax laws were few.

CONVERSION AND REDUCTION OF THE LONG-TERM RATE OF INTEREST

One of the important obstacles to recovery was the existence throughout 1933 and 1934 of a long-term market rate of interest of between 6 and 7 per cent. It hindered private investment and at the same time made it difficult for the government to convert short-term borrowing into long-term. All debtors, but especially public authorities, operated under the burden of high interest payments on past loans. During Bruening's administration enforced conversions of 1931 lowered interest charges on all classes of bonds by at least 2 per cent.

The Hitler government, bearing in mind the experience of 1931, set to work cautiously declaring its program of action would be as follows: "Restore confidence in the validity of contracts by making the state and the economic system secure against disturbing influences from the outside; raise the value of property and increase investment activity by strengthening earnings; and then with these preliminary conditions established reduce the nominal rate of interest." [1]

Outside influences were controlled by means of the foreign exchange control already described, and the government's public works program presumably covered the second stage, outlined in the preceding paragraph. These measures provided a basis for the appreciable rise in the values of both bonds and stocks up to April 1933.

After April, however, a considerable setback occurred in the capital market, and this setback made any possibility of a quick upward movement of stock and bond prices to their nominal values seem most unlikely. Under these circumstances the

[1] I.f.K., *Weekly Report*, Supplement 51/52, 1936, p. 4.

financing of investment by the issue of special bills, which had previously been used only as an expedient, was elevated to a leading principle.

The recession in the stock market in April 1933 may be looked upon as a vote of "no confidence." Lack of confidence occurred at this time because of political events. Hitler had been appointed chancellor on January 30, 1933, and on March 24 the Reichstag with its working majority of National Socialists passed the Enabling Act which brought an end to its own power. The so-called legal continuity which the Enabling Act provided was largely the result of pressure from conservative groups in the cabinet. Despite the ordinary legislative means at their disposal, however, the National Socialists organized riots throughout Germany. For many years the radical elite of Hitler's following had been drilled to think in terms of violent overthrow, and the time had come to satisfy their hunger for action. The violence during the last week in March strengthened the impression that leadership of the country was entrusted to Hitler by the will of the people.

The uprisings also intimidated conservative elements, and hence in good part explained the collapse of the stock market. In this situation, the government refused to decree a general reduction in interest rates or a forced conversion. According to official statement, it was feared that a forced conversion "would disturb the confidence of the public in the securities market."

Throughout the summer of 1933 bond and share prices continued to be depressed. Contributing to the depression in the issuance of new domestic securities was the fact that repayment of domestic bank credits as well as repurchases of foreign securities absorbed most of the available market funds. The unfavorable earnings positions of the banks formed a barrier to reduction of bank interest rates at the same time that repatriation of securities at low prices enabled buyers to make a profit which they were loath to forego.

Another obstacle to voluntary conversion or to a fall in the long-run rate of interest, strangely enough, was the rapid growth of savings deposits. The savings banks were prevented from using deposits to purchase bonds by the banking laws. The Nazis in the Banking Act of 1934, to be described later, made it possible for savings banks to support the government bond market.

Growth of savings demonstrated that savers preferred, on balance, savings accounts to the ownership of securities. Savings deposits yielded only about 3 to 4 per cent as against 7 to 8 per cent for bond yields, but had the advantage, which is not to be underestimated in times of strong fluctuations in security prices, that deposits could be realized without any diminution of value within a short time.

Offering further competition to long-term obligations were other short-term securities such as bank acceptances, treasury bills, or employment creation bills (special government bills). The yields on these securities were, it is true, lower than on bonds, but they had the advantage of being realizable at any time without loss. Finally, there were special securities such as tax credit certificates and very speculative paper, such as, for example, the so-called *Neubesitzanleihe* (pre-war and war loans acquired after 1920 and revalorized at a lower rate than those acquired before 1920).

The first conversion did not occur until 1933, when the Municipal Debt Consolidation and Conversion Act was passed. Conversion was limited to the short-term debt of local authorities. In this way the sum of 2,750 million RM was converted into 4 per cent bonds guaranteed by the Reich. The creditor was not absolutely compelled to accept conversion, but if he refused he was unable to recover either principal or interest for five years, the interest being added to the capital sum at the *old* rate. Steps were also taken in 1933 to reduce agricultural indebtedness by writing down mortgages with assistance from public funds and by permanently reducing mortgage interest rates from their nominal 6 per cent level to 4 per cent.

The bond market was further supported by the law of October 27, 1933, which gave to the Reichsbank new powers to engage in open-market purchases of bonds. Thus the field of activity of the central bank, which previously had been limited to the discounting of bills and making definite kinds of collateral loans, was considerably broadened. By December 1933 the Reichsbank had purchased tax credits from the commercial banks amounting to 350 million RM. Apart from this amount, the Reichsbank made little use of its new powers. Bond prices, nonetheless, rose by an average of 19 per cent from September 1933 to January 1934, while prices in January 1934 already exceeded the low level of April 1933 by approximately 7 per cent.

In 1934 a loan amounting to 329 million RM with interest at 4 per cent was used to convert the small 6 per cent tax-free Hilferding Loan of 1929 and a portion of the old revalorized paper-marks debt of the Reich. Favorable conversion terms were offered, and the saving to the Reich in interest was extremely small.

In spite of improvement in the capital market, a disparity continued between the prices of bonds and securities. On the one hand, the improved earning prospects of industrial companies benefited by the public works program were favorable to stock quotations, while, on the other, the rumors of impending conversions and of possible currency devaluations tended to depress the bond market. The divergent movements of the bond and stock markets led to the question as to how the expected increase in dividends could be advantageously used to stimulate the bond market. The government's answer was the law of March 29, 1934, requiring corporations to invest in bonds.[2] All corporations which declared a higher dividend for 1933-34 than the previous year were required to invest the increase in public loans so far as the declared dividend exceeded 6 per cent. Since only 109 of the 585 stocks traded on the Berlin stock exchange on March 31, 1934, yielded a dividend exceeding the maximum, and since the law did not concern corporations which had reduced capital stock during the previous three years, its effect on the price of public bonds was negligible.

The Loan Stock Act — It is even doubtful whether the further extension of these provisions under the Loan Stock Act of December 1934[3] had any important ramifications until the economy approached the level of full employment. According to the revision in December 1934, all corporations with dividends in excess of 6 per cent, in some cases 8 per cent, were required to pay "surplus" dividends to the Gold Discount Bank, which in turn invested the funds in public bonds. Dividends thus invested in government bonds were to be distributed in cash in 1938, and the law had a time limit of three years. On December 9, 1937, however, the act was extended for another three years, and provisions for cash payments were modified. According to the modification, disbursement of accumulated dividends was made during the early spring of 1938 in the form of non-interest-bearing tax certificates acceptable in payment of taxes during the fiscal years 1941 to 1945.[4] Although these certificates represented negotiable instruments, admitted for trading at the Berlin bourse, shareholders will suffer a loss through the substitution of non-interest-bearing tax certificates for the original interest-bearing bonds.

During 1939 proposals to lower the maximum dividend to 3 per cent were rejected.[5] The act was extended for another

[2] *Rgbl.*, 1934, I, 294.
[3] League of Nations, *Monetary Review*, 1937/38, p. 56.
[4] Guillebaud, *op. cit.*, p. 77.   [5] *D.V.*, September 22, 1939.

three years with the maximum limit still at 6 per cent. Officials stated that the capital market must be carefully nursed with a view to future loans. Suggestions for a capital levy were termed "inappropriate." [6]

The purpose of these various dividend limitation laws was not, as is so often stated, to expropriate the profits of enterprise. Their purpose was to facilitate government financing, and eliminate the influence of the stock exchange as a potential disturbing factor in the economy by stimulating firms to finance capital requirements out of their own profits. Moreover, the law was set up in such a way as to limit plant expansion to industries most strongly influenced by the expanding impact of armaments. The reason for this was that civilian plants could not obtain funds. They were not permitted to float new issues in the stock market, nor were they making large profits which they could plough back into the business. On the other hand, armament firms might avoid the Loan Stock Act by using undeclared profits for expansion and self-financing. In so far as corporate management was reluctant to expand, if "excess" profits were being made, "surplus" income had to be paid to the government, which could then use the funds for any needed expansion. The act also had the effect of making 6 per cent the standard dividend rates, and failure to pay the standard rate depressed stock prices for various corporations, for the most part engaged in non-military production, and made it even more difficult for them to secure funds. The consequent low share prices thus discouraged evasion of governmental restrictions on firms not essential to the war economy.

A decree of June 1941 slightly modified the Dividend Limitation Law, although not in any important aspect. According to Minister Funk, high nominal dividends, even if used in part to buy government bonds, were considered highly undesirable in times of war. The new decree, therefore, heavily taxes divi-

[6] *D.V.*, September 9, 1939.

dends in excess of 6 per cent — the tax is graduated and becomes prohibitive on dividends above 8 per cent. At the same time, however, corporations were permitted to increase capitalization either by issuing new securities or by increasing the face value of existing securities. In this way the nominal dividend rate might be reduced without reducing the total amount of profits distributed among shareholders. Thus the decree is important from a political point of view only, since it prevents charges of war profiteering while shareholders receive the same amount of profits.

During the first three years of the act, only 175 corporations made contributions to the Loan Stock, while the total amount of the stock did not exceed 90 million RM. Considering that consolidation loans issued by the Reich and other public bodies during the same three-year period amounted to more than 6 billion RM, the Loan Stock fund was very small. The largest individual contributor was the Reichsbank, which with its capitalization of 150 million RM and a 12 per cent dividend declared for 1934–37, paid 12 million RM into the fund.

If the first effects of the Loan Stock Act on the reduction of the long-term rate of interest were small, other factors were at work in this direction. Gradually, during 1933 and 1934, the decline in the short-term interest rates, by means of the Reichsbank lowering the rediscount rate, brought with it a decline in bond yields. Moreover, public confidence in the ultimate financial solvency of borrowers was gradually restored. The price of all 6 per cent fixed-interest bonds quoted on the Berlin Stock Exchange had climbed nearly to par in January 1935. Under these circumstances the government decided to convert.

Thus, in 1935, 6 billion RM of mortgage bonds and 2 billion of municipal obligations were offered for conversion at par with 4½ per cent interest instead of the old rate of 6 per cent. For his acceptance of conversion, the creditor was to receive a single payment of 20 per cent of the nominal value of the bonds he held. It was not actually a payment; rather, the amount was deducted from taxes which the creditor owed the government.[7] If creditors did not reply to the notice within ten days, it was to be assumed that they were in agreement with the reduction. If the holder refused conversion, he was permitted to draw the former rates of interest, but his bonds lost their negotiability — they were no longer quoted on the Stock Exchange or eligible as collateral at the Reichsbank.

The response was overwhelming, only a few creditors refusing the conversion. All the forces of propaganda were used while German investors were unable to transfer their money abroad owing to foreign exchange control. In any case the whole manoeuvre was a success, for in November 1935, after the 20 per cent bonus no longer influenced quotations, the price of the bonds stood at 95 as compared with 96 before conversion.[8]

The law of February 27, 1935, using the same ingenious devices, enabled the commercial banks as well as the states and municipalities to reduce interest rates to 4½ per cent. In turn, the savings banks and insurance companies reduced their rates. Finally on July 2, 1936, interest on non-agricultural private mortgages followed suit.[9] The government provided that conversion should proceed "voluntarily." In case of failure to arrive at a decision, the courts were empowered to establish a suitable rate of interest, the rate allowed in such cases being as a rule 5 per cent on first mortgages and 5½ to 6 per cent on second mortgages. Interest rates on industrial loans were left exclusively to voluntary conversion.

No other reduction in the long-term rate of interest was carried out by conversion or decree until April 9, 1940. In the midst of the war the government declared that there should be

[7] Department of Overseas Trade, *op. cit.*, pp. 27–29.
[8] Guillebaud, *op. cit.*, p. 79.
[9] Department of Overseas Trade, *op. cit.*, p. 30.

a decrease of ½ to ⅝ per cent below the previous rate of 3 per cent on savings deposits.

At the beginning of 1941, however, there were indications that the government might reverse its easy-money policy. In an address at Hamburg on February 23, 1941, Vice President Kurt Lange of the Reichsbank announced that the Ministry of Finance and the Reichsbank had no intention of adopting any policy to lower the current standard of 3½ per cent for the nation's medium-term borrowing. According to Lange, further reduction might produce undesirable repercussions in "various sectors of the German economy" — a statement which, it is believed, referred primarily to savings and investment-insurance reserves. The announcement of a halt in the cheap-money policy tended to check the advance in stock prices but was also interpreted as notice of abandonment of any plans for important conversions.[10]

REFORM OF THE BANKING SYSTEM

Following the banking crisis of July 1931, various reform measures were proposed, but the only legal amendments evolved were the creation of the office of a banking commissioner. On December 5, 1934, a new bank and credit law was inaugurated. It was the outcome of an investigation held partly in public and partly in private and of a final report visibly inspired by Dr. Schacht. The plans for reform varied from conservative proposals to complete nationalization. The act of 1934 both in wording and aims avoids revolutionary changes of every sort. Its ultimate objectives had been clearly stated in the final reports as follows:

The only way in which the government can provide itself with the necessary funds is by raising loans. But loans are possible only if the savings of the population are entrusted to the State through the medium of the capital market. The establishment of a money

[10] *New York Times*, February 24, 1941.

and capital market, capable of fulfilling the tasks of the State must be the ultimate aim of any kind of reorganization. All legal measures must be framed with this aim in view, and no organization or institution can be allowed to remain outside the system.[11]

How does the act solve the problem of building up a banking system which is completely at the disposal of the government, but nevertheless follows the fundamental rules of commercial banking under healthy market conditions? The answer is found in the principal provisions covering liquidity, credit facilities, publicity, and control.

Control — A credit supervisory board consisting of the president and vice president of the Reichsbank, four secretaries of states and a member appointed by Hitler, together with a Reich-commissioner for credit was established as a permanent authority. An amendment of June 15, 1939, however, changed this supervisory board to an autonomous Reich control board headed by a separate president and authorized to act as an executive authority under the Minister of Economic Affairs for all questions pertaining to credit policy. The effect of this is to shift the main control from the Reichsbank to the Minister of Economic Affairs and hence signifies even greater centralization than that provided by the act of 1934.[12]

The Board now has power over all banks and credit institutions including the Reichsbank and the Gold Discount Bank. The functions of the Board far exceed those of merely enforcing legislation, and its principal means of control are (1) licensing, (2) supervision of audits, (3) information and sundry powers.

The Board's licensing power enables it to determine the number of new banks and branches, as well as to control amalgamations. Licenses may be refused if the existence of a bank is unjustified with reference to the local economic needs,

[11] Marie Dessauer, "The German Bank Act of 1934," *Review of Economic Studies*, June 1935, pp. 214–215.
[12] I.f.K., *Weekly Report*, December 12, 1939, pp. 97–100.

or if the owner or manager does not possess the necessary qualities of character or training, or if other qualifications and experience are lacking. These rather vague reasons for refusing or withdrawing a license give the government somewhat arbitrary powers of selection.

The Board is entitled to lay down regulations regarding method and scope in auditing, which up to 1933 had been compulsory only for joint stock banks and for mortgage banks.

The president of the Board also has the express right to ask any credit institution at any time for balance sheets and profit and loss accounts and any information which he may think necessary. He is entitled to scrutinize at any time books and accounts of any institution, including statements of the position of assets and liabilities, especially as to foreign exchange. He may take an active part at general meetings, and determine the agenda. In case of urgency he is even entitled to act on behalf of the banks.

The authorities thus have a complete survey of all financial events and may determine particular policies through this specialized knowledge. Moreover, centralization and inclusion of all banks under the control system enables the government to influence a bank's credit policy according to its wishes. Although profits are limited by government decisions, they have been substantial, and are retained by individual owners of the bank.

Credit Regulations — The main regulations in regard to the granting of credits may be classified as follows: (1) size of credit, (2) kind of credit, (3) unsecured credits.

The reason for regulations as to credit limits grew out of abuses in post-war financial policy. Banks frequently gave credits to industrial companies out of all proportion to the firms' assets and to their present or future earning power. Frequently banks made funds available beyond the bank's lending capacity. This form of financing proved feasible in times of considerable technical progress, but was disastrous otherwise. The act thus aimed at preventing mismanagement of this kind.

Credits of a bank to a single firm must not exceed a certain percentage of the bank's own capital — the percentage to be determined by the Supervisory Board. The Supervisory Board has enforced a limit on credits of 5 per cent of the bank's capital. Between 1931 and 1934 the banking commissioner, in case of reconstructions, had already advised this same percentage. With this exception, however, the amount accorded to one firm may be twice the legal percentage if all partners or managing directors of the bank expressly agree, but credits in excess of the legal percentage must be reported to the banking commissioner. In any case, banks must inform the Supervisory Board whenever the total debit of one single customer exceeds 1 million RM in the course of one month. The law includes detailed regulations regarding the definition of a "credit" and of a "firm," chiefly in order to avoid the granting of credits to several companies which should correctly be regarded as the same business.

Compelling information on credits above a certain amount is an old German regulation that appeared shortly after the time of the inflation. It was introduced because of the difficulties created by the fact that different banks gave credits to a single firm without realizing, until it got into difficulties, that the firm had other creditors. Big concerns, until the Nazi period, refused to submit detailed balance sheets even when they asked for large credits.

The new law aims to provide a "neutral" board which can pool the knowledge concerning the bank commitments of all bank clients. In this way, the privacy of the client, considered as essential by commercial enterprises, is protected. The law also gives the government a definite knowledge of credit distribution.

These provisions have the appearance of forcing the banks to distribute credits among a large number of applicants and of restricting the size of credits. Exceptions which may be granted, however, by the vote of the directors of the bank involve considerable leeway. Moreover, credits to government or credits guaranteed by it are excluded from all provisions regarding credit facilities. The more probable consequence of these rules is the direction of credit toward the ends of the state and the efficiently powerful groups. For in so far as the act enforces distribution of credits in small lots and on a rationing basis, it prevents firms from freely expanding capital or setting up new enterprises on the basis of bank facilities. It thus forces them to go to the stock market and hence to obtain government consent for expansion. If the firm's activities come within the scope of the war drive, the permission may be granted. The act, moreover, gives a wide latitude for discretionary powers, so that, as among various armament plants, the one managed by a man in the good graces of the party may be favored.

In order to enforce sound banking principles, total investments in real estate and permanent participation in other companies may not exceed the capital of the bank. Holdings of stocks (except those representing permanent participation) and of securities not quoted on German stock exchanges are limited to a certain percentage of total liabilities less savings deposits. The Board from time to time determines the percentage.

An additional regulation proposes to safeguard the banker and at the same time prevent new entrepreneurs from undertaking investments which have no favorable prospects. In the case of unsecured credits of more than 5,000 RM, the bank may demand from the firm a detailed balance sheet and also may require adjustments for reasons of "sound" financing.

Regulations of credit grants to business have not prevented a considerable rise in bill circulation. The increased note circulation has been looked upon by some investigators as a sure sign of an unhealthy financial condition. Expansion of note circulation ceases to have much significance when prices and wages are regulated, the government has a monopoly of the capital market, and propaganda has created considerable enthusiasm on the part of individual savers for government loans. The activity of individual savers may also be buttressed by the considerable trend toward institutional savings. Although the note circulation at the end of 1940, amounting to slightly over 14 billion RM, was greater than at any time since the reorganization of the German banking system after a runaway inflation, the additional purchasing power was tapped by new taxes and borrowing.[13] If the people once lose enthusiasm for state loans, increased purchasing power requires additional taxes or a rise in prices. In February 1941 bill circulation dropped to slightly over 13 billion RM, despite the fact that there had been no change in the discount rate. In the face of the considerable rise in bill circulation the Reichsbank had lowered its rate from 4 per cent to 3½ per cent in April 1940. Continued support of government loans maintained a ratio of 50–50 between tax revenue and loans until March 1940, and after that time there was even a slight preponderance of loans over taxation.[14]

Liquidity and Reserve Regulations — Of great practical importance is the power of the Board to determine various bank reserves. The Board prescribes two different reserves against total liabilities (other than savings deposits). The first line of reserves, called the cash ratio, determines the amount of cash kept on hand against all liabilities excepting savings deposits. It is fixed by the Board from time to time but, in order to protect earning power, can never be set higher than 10 per cent. The second line of reserves, consisting of 90-day commercial bills, Treasury bonds, and public loans, is likewise determined by the Board, subject to the limitation that the rate may never

[13] *New York Times*, February 27, 1941.
[14] *New York Times*, October 28, 1940.

go above 30 per cent. The final reports of the act emphasize the desirability of keeping the assets required for the secondary reserve in Reich loans and in the loans of states and municipalities.

Savings and coöperative banks so far as they carry other than savings accounts are also subject to these reserve regulations. Earlier savings bank regulations were even more severe as to amount of coverage on total liabilities, but reserves in the case of the savings banks could be deposited at the regional or central clearing associations or, in the case of the coöperative institutions, at the central coöperative bank.[15] Under the 1934 act, reserve funds must be retained at the bank. In order to make bigger profits, the banks have preferred to keep the reserves in interest-bearing government bonds, eligible as assets for the secondary reserve, rather than to retain them in cash which bears no interest. In this way the considerable growth in savings deposits has been used to support government financing.

Credit institutions must invest savings deposits separately and likewise keep separate accounts for such transactions. Premature withdrawal of large sums in savings deposits is discouraged. Deposits for the first four weeks earn no interest, and funds which by their very nature are available only for a short time must be paid into transfer accounts bearing a small or zero interest rate, as in the American checking-account system.

Under the aegis of the Nazi Banking Act of 1934, the commercial banks actively assisted the government in placing long-term loans. The banks used their whole equipment and business connections in the services of the Nazi credit policy, and it has more than once been explicitly and officially recognized that the active coöperation of the banks has been a vital factor in insuring the success of the funding operations. The banks themselves took up these loans so far as their own re-

[15] *Rgbl.*, 1933, I, 103, 1080.

sources and considerations of liquidity permitted them to do so.[16]

The Banking Act is, moreover, important in obtaining not only an active market for bonds but also uniformity of banking facilities. Special difficulties, however, are encountered by private banks. According to the law they must follow exactly the methods of business of the joint stock banks, and it is therefore to be feared that their good will, as far as it consists of personal experience and close commercial connection with certain branches of industry and of a readiness to take an active interest in their clients' affairs, will be of little value.

Since the law does not provide for governmental interference in the day-to-day activities of the banks, it gives the impression that each bank, private as well as joint stock companies, has a wide scope for private initiative, and it leaves the public with the feeling of being served by an individual banker and not by an official authority controlled by powerful groups. According to one of the managing directors of the Reichs-Kredit-Gesellschaft, O. Fisher, "the authorities are anxious to maintain the appearance of private initiative because it is feared that the public might discover in the nationalization of banking the right of the government to an unconditional claim on credit facilities." [17]

THE STOCK EXCHANGE ACT

The stock exchanges were reorganized by the Nazis according to a decree of March 4, 1934, but no revolutionary changes were inaugurated. Previous to the 1934 decree a German stock exchange had to obtain a charter from a state government, but direct control of the mechanism of trading was left to the relevant Chamber of Commerce. Under the Nazis, the charter

---

[16] Reichs-Kredit-Gesellschaft, *Germany's Economic Situation in the Middle of 1938*, p. 92.

[17] Marie Dessauer, *loc. cit.*, p. 224.

is obtained from the Reich rather than the state. All members of the exchange board as well as its president are still named by the Chambers of Commerce, which still exercise direct control over the exchange. The president of the exchange board makes all important decisions; other members of the board have merely an advisory function. The fixing of quotations is done by official brokers whom the state governments appoint and recall — in Prussia, for example, the Prussian Minister of Economic Affairs appoints him.

Certain activities which led to abuses were prohibited during the banking crash of 1931. Since 1931 all stock transactions have been for cash only. Before that time, security loans were divorced from the trading market — made directly by the banks. In 1936 additional measures regulated the use of blocked balances. Undesired fluctuations sometimes occurred because blocked balances were used to buy stocks or because foreign holders of German bonds sold them and bought stocks. From the point of view of conversion and consolidation of public short-term credits, a diversion from the bond market to the stock market was by no means desired. Therefore, in an act of December 19, 1936, it was decreed that shares could be purchased with block balances only if the blocked balances had arisen from sales of stocks. Thus, after that date, stocks could only be acquired in exchange for stocks. On the other hand, blocked balances might be used freely to buy bonds.[18]

A corporation dealing in its own shares in the stock market had been frowned upon since 1931, but no special provisions to cover this activity were made by the Nazis. As in 1931, all transactions of this type must be included in the annual report of the company. Nor has any provision been made prohibiting insiders from rigging the market. The new exchange act, as in earlier examples, denies the obligation and possibility of protecting individuals in security trading activities. Official regu-

[18] *Rgbl.*, 1936, I, 189.

lation of speculative credit is apparently nonexistent, and a great deal of reliance is placed on the principle of "caveat emptor."

A significant reduction in the number of German stock exchanges was required under the law, a reduction from 21 to 9. Provisions for admission of issues for trading on the exchange were made more severe to prevent random fluctuations in what had become a very limited market. Stocks issues listed on the Berlin exchange must have a nominal value of at least 1,500,000 RM; the former minimum was only 500,000 RM. On the Frankfurt and Hamburg exchanges a minimum circulation of 500,000 RM is necessary before the stock is listed; for the other exchanges the minimum remains as it was, 250,000 RM. Issues of less than 3 million RM which are closely held and relatively inactive may also be excluded from the Berlin exchange and be forced to find listing on another market. The act is thus biased in the direction of protecting the negotiability of very large issues.

So far as new issues are concerned, the federal government permits only its own or only such others as serve the purposes of the war economy. This monopoly has meant a considerable reduction in the business of the exchange, as can be seen from the decline in the exchange sales tax. Inasmuch as the rate of the tax has not changed since 1925, the decline, indicated in Table 11, fairly accurately represents the diminished importance of the securities market. Smaller savers are placing their money in life insurance and savings banks because under the Corporation Law of 1937, all stock issues are too large, 1,000 RM, to be available to them. Financing of small concerns is carried on by the banks outside of the stock exchanges.

Already under the Weimar Reich savings and investment habits altered, for the activities of security markets in industrial financing "decreased, when the concentration of industrial enterprises increased the importance of self-financing and favored stock turnover outside the markets."[19] The Loan Stock Law further encouraged self-financing. Social insurance had also caused fundamental shifting in investment habits, because it undertook the provision for old age which was formerly made largely by purchase of bonds.

TABLE 11
STOCK EXCHANGE SALES TAX RECEIPTS
(*1,000 RM*)

Year	Amount
1925	62,786
1926	61,955
1927	83,338
1928	49,359
1929	33,188
1930	21,198
1931	13,720
1932	8,298
1933	11,480
1934	13,681
1935	14,494
1936	15,833
1937	16,699

SOURCE: I.f.K., *Weekly Report*, Supplement, March 9, 1938.

Is the stock exchange, then, nothing but a meaningless carcass without function? An interesting answer to this question is contained in the *Weekly Report* of the German Institute for Business Research: "Even though the evaluation of securities is no longer under the influence of the free play of forces, it is still of great importance. For the individual still has a basis for evaluating his fortune and for the government, bond yields show to what extent measures in this field have been successful or to what extent they must be changed."[20] Security markets assure shareholders of the possibility of selling their shares. With the principle of negotiability stands or falls the principle of financing by security issues, "which principle the Nazi Gov-

[19] I.f.K., *Weekly Report*, Supplement, April 1937.
[20] I.f.K., *Weekly Report*, March 9, 1938.

ernment is not willing to abandon." [21] Negotiability is important in new issues no matter whether these issues are placed directly through exchanges or outside of them, for the fact that there is a stock market where stocks may be exchanged adds to the negotiability of shares not traded on the market.

DEFICIT FINANCING

Public works and later armament needs were financed by special bills which represented short-term borrowing on the part of the government. The volume of such bills was kept secret in order to prevent other countries from realizing the strength of Germany's long years of rearmament effort. What was clear, however, was that the amounts were unprecedented. New and unusual techniques were devised to prevent short-term financing from aggravating the problem of price and wage controls. The government was always hypersensitive to inflation dangers, and in 1935, after successful conversion had reduced the long-run rate of interest from 6 to 4½ per cent, it successfully floated a series of funding issues which consolidated the short-term debt.

The government's financial task was facilitated by the fact that not all of the increased national income was available for consumption. Taxes and contributions described in the next section took a greater share than before, and voluntary savings increased. Savings appear to have increased more than the amount normally to be expected with a rise in income. The increased rate of saving may be explained by the increasing inequality of income, the shortage and rationing of consumers' goods. Moreover, financial policies lose much of their significance when raw materials and labor are strictly regimented.

As in other fields, German financial policies did not follow a preconceived program. On the contrary, each emergency situ-

[21] *Ibid.*

ation was met as it appeared, and under no circumstances were preconceived ideas allowed to interfere with the progress of military expansion. When rearmament started, it was necessary to use short-term bills in order to allay the fears of illiquidity in the banking system. Devices were discovered, however, which made it possible to produce armaments without being hampered by the alleged impossibility of increasing the government debt.

The various devices used in important emergency situations will next be described. When the volume of special bills was still fairly small, prior to October 1933, they were generally discounted by ordinary commercial banks, which at that time were glad to find a safe investment. In October, however, the banks' portfolios were filled to overflowing with special government bills, and they were forced to rediscount part of them with the Reichsbank. The Reichsbank intervened to prevent the rise in the market rate of discount by discounting the bills on a large scale directly. At the same time, it used newly created open-market powers to buy up tax certificates, thus lowering the discount at which these certificates had been circulating.

From June 1933 to December 1934 the banking system was highly illiquid and heavily in debt to the Reichsbank; consequently the special bills were used to reduce indebtedness of business at commercial banks and of commercial banks at the Reichsbank. On balance, there was probably not much change in the total volume of credit outstanding. With the restoration of liquidity, short-term interest rates fell from an average of 5 per cent in 1933 to 4½ per cent in December 1934. Bank advances still carried interest rates as high as 7 per cent, but business could obtain credit by discounting special bills. In view of the fact that the Reichsbank was using special bills to deplete cash reserves (income, business, and precautionary deposits), there was no necessity for the velocity of circulation or the short-term rate of interest to rise as a result of the rise in incomes.

The emergence of excessive cash reserves, moreover, was prevented by a special form of market operations resembling the policy adopted in England during the first World War. When large funds began to accumulate in the banks and the Reichsbank's portfolio was heavily weighted with these special bills, the Gold Discount Bank, the "daughter" institution of the central bank, stepped into the market and bought government bills from the banks, offering its own promissory notes in exchange. These promissory notes represented a liquid security for the banks and at the same time relieved the central bank in financing state projects; the Gold Discount Bank applied the funds received from other banks to take over work-creation bills held by the Reichsbank. Since these funds represented the surpluses and reserves of industrial and other enterprises which benefited especially from state orders, a circular process of funds had completed an orbit from central source to production system, thence to the money market, and again to the central bank and its subsidiary.

Funding Operations — However, this was but an intermediate stage of the financing process for the public works and armament projects. The last stage began with the funding of the short-term credits by means of long-term loans. In January 1935 the first step was taken toward consolidation of the floating obligations of the Reich arising from the special bills; a loan of 500 million RM was issued at 4½ per cent to the savings banks. The subscriptions were made by installments extending to August 1935, and the liquidity rules of the savings banks were relaxed in order to facilitate the absorption of the issue. As savings deposits had increased considerably in 1934, the savings banks were able to provide the installments with only temporary recourse to the money market.

In May arrangements were made for the taking up of a similar loan by insurance companies. The fact of its conclusion was never officially announced; the amount actually raised was not divulged, but it appears that by the end of the year some 300 million RM had been taken up in monthly installments. This loan was made possible through the sharp rise in life insurance contributions. Increased premium payments and a decline in surrenders and borrowing on policies had added considerably to the sources of capital accumulations.

One-half of the second funding loan (1 billion RM), issued in September 1935, was offered directly to the public. At shorter intervals other loans followed, among them a 500 million RM loan floated by the Railways Administration to redeem the bills by means of which the construction of highways for automobiles had been temporarily financed.

One of the peculiarities of these transactions was that to a large extent they were not carried on through the open market, the new securities being taken up directly by the most important agencies of capital accumulation, the savings banks and insurance companies. However, a further important stage in the process of liquidation was reached in June 1936, with the issue of a government loan amounting to 700 million RM, the main part of which was offered for subscription on the open market. By the end of September 1940 the amount of the various funding loans totaled over 36 billion RM.[22]

In 1936 deposits in commercial banks for the first time began an upward movement reflecting the need for replacement of plant and new industrial equipment after more than three years of continuous and intensive utilization of the industrial facilities. The considerable increase in the total of Reichsbank giro clearings, the best index of total volume and turnover of banking credit, together with the growing inelasticity of the short-period supply curves of industrial production pointed the way

[22] League of Nations, *Monthly Bulletin of Statistics*, XXI (November 1940), 440.

to an important change in financial policy in the spring of 1938.

After April 1, 1938, the era of what the Germans call "prefinancing" was scheduled to end. At that time, it was declared that only tax revenue and long-term issues on the capital market were to be used for financing new investment. To provide for the period of transition from the old to the new, the Minister of Finance was empowered to issue non-rediscountable bills having six-months' maturity and bearing only 3 per cent. These were to be funded by periodic long-term issues, and were known as delivery bills.

The objective of the new policies was to enforce economy so that more direct controls over production and consumption would not have to be introduced. Economy, however, was never achieved, for occupation of Czechoslovakia and hurried construction on the West Wall brought additional increases in the floating debt.

In the spring of 1939 a program known as the New Finance Plan was adopted. New types of tax redemption certificates took the place of delivery bills. All public purchases to the extent of 40 per cent were paid by these certificates. Each person who received such payment could in turn use the certificates for as much as 40 per cent of business purchases. This financial instrument, however, proved so cumbersome and complicated that the issue of tax redemption certificates was suspended as from November 1, 1940. After the war broke out, the government financed itself in the traditional manner — by treasury bills, notes, and bonds. In addition, special measures provided for war problems.

War Measures — In connection with the war it should be pointed out that little change was demanded. Control of the money markets and capital markets had already been centralized and simplified by six years' experience at war financing. Moreover, it was no longer a revolutionary idea to have a large volume of private funds pass through the "public hand."

Specific war measures were related to activities of the "Oeffa" Bank, the Deutsche-Industrie Bank, and the Reichs-Kredit-Gesellschaft. Special war bills have also been issued.

The "Oeffa" Bank was empowered to extend short-term credits, called certificates for military needs, to firms whose liquidity position was seriously hampered by the war. The "Oeffa" may grant these credits only if all normal credits have been exhausted, if the additional credit needed are exclusively a result of the war, and if the credit provided will be sufficient to remove the difficulty. The Deutsche-Industrie Bank is largely concerned with investment credits needed in reorganization to meet war demands. It grants credits only if a company's own bank is unwilling to undertake the financing because of risk or lack of funds. The applicant's own bank acts as intermediary. Current credits or investment credits for reorganizations for companies in foreign trade are handled by the Reichs-Kredit-Gesellschaft. The company applies for the credit through its own bank, and the Reichs-Kredit-Gesellschaft guarantees the loan. Special war bills or certificates for military needs, 60- to 90-day bills, are created for very special demands caused by shifts to war production. These bills are discountable at a commercial bank, but cannot be deposited with the Reichsbank. They thus have limited expansion.[23]

GOVERNMENT DEBT AND REVENUE

The advent of Hitler brought with it the slow and efficient mobilization of all Reich finances. To the same degree as the armed forces, the nation's fiscal machinery was geared for military activity, so that at the outbreak of war one of the most impressive German performances was the clocklike regularity of the fiscal machinery. Apart from a few new taxes imposed in September 1939 hardly any major tax increases or adjustments were required. The rising level of government debt

[23] I.f.K., *Weekly Report*, September 5, 1940.

created little trouble, because the government controlled the stock market, banking regulations were centralized and unified, and prices and wages were fixed.

According to Hitler, Germany spent 90 billion RM for war preparation, while the latest available estimate shows that during the first year of actual war Germany expended another 50 billion RM. Dr. Fritz Reinhardt, Assistant Finance Minister, has indicated that the outlay for the fiscal year 1940–41 has probably been at the rate of 5 billion RM monthly.[24] How have these large expenditures been financed? How was war financing carried out with rigid precision?

To answer these questions, it is necessary to analyze first the tax system and then the debt structure. A war tax system was handed to the Nazis "ready-made"; they merely continued the high rates of the period of deflation. Despite the fact that they adopted an easy-money policy, and a program of state deficit-expenditure, the regressive system of taxes was not changed to ease the burden on the consumer. With the war boom, increased incomes brought increased tax revenue but little alteration in the share of national income going to taxes. In 1937 taxes as a percentage of total national income amounted to 26 per cent as compared to 25.5 per cent in 1932. The slight increase may be attributed to increasing severity of the measures against tax evasion. In the previous prosperity, 1929, the ratio came to only 17.5 per cent.

A brief description of various taxes, together with statistics showing tax payments falling largely on property and those borne by labor, will demonstrate the unusually regressive nature of the tax system. As we shall see, increased corporation taxes had modified this picture by 1939. The increase in the income tax immediately following the outbreak of war additionally increased the share of taxes on the upper income groups. This trend was, obviously, inevitable. With the high rates already

[24] *New York Times*, August 26, 1940.

enforced on the lower brackets and wage rates remaining at depression level, the wage tax could hardly be a source of further revenue. Other taxes bearing most heavily on labor were also compelled to fall off — namely excises, duties, and turnover taxes. This was so because of the blockade and the rationing of consumers' goods. Textile rationing alone brought a reduction of 5 billion RM which before the war would have been spent on clothes, and hence have increased the revenue from turnover taxes.[25] The rate of increase in income, moreover, was largest for corporations and upper income brackets, and it was necessary that their share of the tax burden be increased.

A summary description of changes in various tax rates will reveal the nonprogressive nature of the tax system which was taken over and maintained by the Nazis.

Income Tax — From 1928 until the War Economy Decree of September 1939 income tax rates ranged from an average of 10 per cent on income below 8,000 RM to an average of 40 per cent on the portion of income above that amount. Income of 1,300 RM is exempt from the tax.[26] Various measures involving changes in allowable deductions, increased tax rates for bachelors, inclusion of servants in family allowances, and other similar palliatives during 1930–34 did not appreciably alter the character of these tax rates. Income taxes as a percentage of total taxable income increased from 10.9 per cent to 13.6 per cent in 1936, the latest year for which this detailed breakdown is available. Additional details concerning the burden of the income tax according to income classes are presented in Table 12.

The War Economy Decrees raised the tax rates by 50 per cent on all incomes over 2,400 RM but set a maximum of 65

[25] *New York Times*, February 8, 1940.
[26] League of Nations, *Public Finance, 1928–35: XII (Germany)*. (Geneva, 1936–39), p. 7.

per cent. This meant, for example, that whereas an income receiver of 5,000 RM paid about 10 per cent before the Decree, he would pay 15 per cent after the decree. In view of the fact that the increase was the same percentage for all income classes with a limit for the highest rate, it represented a significantly increased burden for the middle income groups. As tax rates

TABLE 12

DEVELOPMENT OF THE BURDEN OF GERMAN INCOME TAX ACCORDING TO INCOME CLASSES

(*Amount of tax paid in each class interval as per cent of taxable income received by that class*)

Income Classes (RM)	1928	1932	1933	1934	1935	1936
Under 1,500	3.0	4.2	4.3	3.1	3.2	3.2
1,500 to 3,000	4.5	5.4	5.4	4.9	5.1	5.1
3,000 to 5,000	6.0	7.1	7.1	7.2	7.4	7.3
5,000 to 8,000	6.9	8.3	8.2	9.1	9.3	9.2
8,000 to 12,000	7.9	9.2	9.2	10.8	11.0	10.9
12,000 to 16,000	9.4	10.7	10.7	12.7	12.8	12.8
16,000 to 25,000	16.9	13.9	13.8	15.6	15.6	15.6
25,000 to 50,000	17.6	20.4	20.3	22.0	22.1	22.1
50,000 to 100,000	25.2	28.3	28.0	29.4	29.5	29.0
100,000 and more	34.1	37.3	36.0	37.5	35.6	30.8
Total tax as per cent of total income received	10.9	11.6	11.3	12.1	13.0	13.6

Compiled from data in *W. und S.*

increased, the principle of progression takes on a more important role; it was a principle disregarded in this case. On the other hand, the burden which the very largest income groups had was large; an individual who earned a million RM in a year might have a mere 350,000 RM left to spend and save. A middle-income receiver with 5,000 RM would have 4,250 RM after deduction of the war income tax.

Corporation Tax — From 1928 to August 27, 1936, the corporation tax amounted to 20 per cent of corporate net income. The law of 1936 raised the tax to 30 per cent, and again on July 25, 1938, the tax on corporations with an income in excess of 100,000 RM increased to 35 per cent. The same law provided for a tax of 40 per cent for 1939 and 1940, and the War Economy Decree did not raise the rate.

Property Tax — In 1929–30 a tax of 8 per cent was levied with a property exemption of 5,000 RM. In January 1931 the exemption was increased to 20,000 RM. Moreover, rates were lowered to 3½ per cent on December 1930 and again to 3 per cent for 1932 and 1933. They were raised again on October 16, 1934, to 5 per cent — no change in the exemption.[27] The present rate is thus below that of the previous boom but slightly above that for depression. All proposals for a capital levy at the outbreak of the war were rejected.

Inheritance Tax — There has been only one change in the only slightly progressive inheritance tax rates. This was a small modification on October 16, 1934, which allowed a larger amount of tax-free property if the heir was closely related to the deceased person. Rates vary according to the relationship between the deceased person and his heir or legatee and according to the magnitude of the amount received. The average rate for bequests up to 10,000 RM is 8 per cent, while the average rate for bequests of more than 10 million RM is 37 per cent. The law is clearly less progressive than the American rates.[28]

Tax on Wages and Salaries — The tax rate on earned income below 120,000 RM has had a varying course in the period from 1928 to 1935, the last year in which there was a change in the law. The most important revisions occurred in the laws of March 24, 1934,[29] and October 16, 1934.[30] These revisions were aimed at increasing the family allowance and consolidating the marriage and unemployment contributions with the

[27] League of Nations, *Public Finance, op. cit.*, p. 7.
[28] League of Nations, *Public Finance, op. cit.*, p. 7.
[29] *Rgbl.*, 1934, I, 235.
[30] *Rgbl.*, 1934, I, 1005.

earned income tax. The rates in each year vary considerably according both to the size of income and the size of family.

As is shown in Table 13, the rate for a single man was considerably higher in 1935 than in 1928. However, for a married man with two children the rates were less if income was low, although higher for incomes above 4,800 RM. The rate in 1928

TABLE 13

TAX PAYMENTS OF WAGE AND SALARIED WORKERS IN PER CENT OF EARNED INCOME

	UNMARRIED			MARRIED WITH 2 CHILDREN		
	Per cent			Per cent		
Income (RM)	1928	1935	1928=100	1928	1935	1928=100
1,200	..	2	(no tax 1928)	..	..	...
1,800	3	6	221	..	..	...
2,400	4	9	223	2	2	84
3,600	6	14	330	4	4	95
4,800	6	15	211	5	4	91½
6,000	7	15	208	5	5	101
7,200	8	17	213	5½	7	121
9,600	9	18	212	6	8	133
12,000	9	20	215	7	9½	138
18,000	11	25	221	9	13	142
24,000	14	30	211	12	16	137
36,000	19	38	204	17	22	130
60,000	24	47	192	23	28	121
120,000	31	50	160	30½	40	131

SOURCE: *Einzel.*, 1937, Nr. 35, p. 141.

progresses from 3 per cent on 1,800 RM to 31 per cent on income of 120,000 RM. The 1935 law begins at 1,200 RM with a rate of 2 per cent and progresses to 40 per cent for the highest salary class. In 1935 the rate of increase in tax for both single and married men was greatest for the middle income classes.

Turnover Tax — The turnover tax was raised during 1932 to 2 per cent (as compared to 0.75 per cent in 1928) and was never reduced except on farm produce, where it was lowered to 1 per cent on October 2, 1933.[31]

[31] League of Nations, *Public Finance, op. cit.*, pp. 8–9.

156   THE STRUCTURE OF THE NAZI ECONOMY

Excise — Excise duties on cigarettes and cigars were increased in December 1930, again in December 1934, and still again under the War Economy Decree. Excise duties on sugar were doubled in June 1931, while duties on beer rose only as a result of the war. An increase in the price of alcohol, long a government monopoly, was introduced in October 1935 and again during the war. New receipts were created as from May 1930 by the introduction of duties on mineral oil, and the duty on salt which had been abolished in April 1926 was introduced again in June 1932. In April 1933 a tax on fats — margarine, artificial fats, and alimentary oil — was levied at the rate of 50 RM per kilogram. A tax on slaughtering appeared as of May 1934 with rates varying according to the animal slaughtered and weight — from 4 RM to 22 RM per head of cattle.[32] Duties on sparkling wines, which were suppressed as from December 1933, were not reintroduced until the War Economy Decree.

Customs duties were, of course, subject to changing decrees of the foreign trade control boards. The yield showed a steady decrease, not because duties fell but because the control board rationed imports.

Revenues in General — The continuation of high tax rates bore fruit with the return of prosperity, and revenues from Reich, states, and municipalities rose from 13½ billion RM to 22 billion in 1939 and 27 billion in 1940. Total public revenues during six years of Hitler's regime before the war were 35 billion RM higher than they would have been if revenues had remained at the level of 1932–33.

Thus the government was able to meet a large proportion of its financial needs from current revenue, for until the outbreak of war a ratio of at least 50–50 between tax revenue and loans was the government's recipe for financing war preparation (see Table 15).

[32] League of Nations, *Public Finance, op. cit.*, p. 9.

Aside from tax revenue, the government had available the amount saved in unemployment relief expenditure. As more people were put back to work, the cost of unemployment assistance declined from over 3 billion RM in 1932–33 to an estimated 300 million RM for 1938–39.[33] These savings totaled over 10 billion RM by April 1939.[34] By converting a substantial part of the debt as early as 1934–35, the government economized on interest charges. Not included in the statistics of government revenue was a levy on business for subsidizing exports. The amount of the levy is unknown, but it is quite generally believed that it has amounted to one billion RM annually since 1935.

Government Debt — With public expenditures growing, the current resources of the government had still to be supplemented by large-scale borrowing. Although the short-term debt continued to increase, the government also floated a large number of medium- and long-term loans. The entire public debt, including the secret debt, amounted to 59.1 billion RM in March 1939, and 77.3 billion RM in 1940, as compared to 24.3 billion RM in March 1933. The secret debt represents the amounts of bills drawn by contractors of government orders. These bills were not markedly different from ordinary commercial paper. They could be discounted at savings and credit banks, sold on the money market, and were rediscountable at the Reichsbank. The government preferred, however, to cling to the fiction that these special bills did not represent short-term debt.

The various categories of the entire public debt is presented in Table 14. The amount of the debt which is held at short term (declared and concealed) does not appear to be excessive, be-

[33] Otto Barberino, "Veraenderte Struktur des oeffentlichen Haushalts," *Der Wirtschaftsring*, July 22, 1938 and *W. und S.* XIX (1939), 575.

[34] Savings on unemployment assistance is computed by deducting expenditure in any given year from the amount spent in 1932–33.

cause of the government's success in continually funding the short-term debt. Moreover the size of the public debt is not alarmingly large in relation to national income. Comparisons

TABLE 14

The Debt of the Reich, State, and Municipalities *

(Billion RM)

As of March 31	(1) Total Debt (cols. 2, 3, 4, 5, 7)	(2) Foreign and Domestic State and Municipal Debt †	(3) Foreign Debt ‡ (Reich)	(4) Old Domestic Debt (Reich)	New Domestic Debt Reich §		
					(5) Total Declared	(6) Short Term	(7) Secret Debt
1929	14.7	7.5	0.9	5.6	0.7	0.2	..
1930	20.3	11.7	1.1	4.9	3.6	1.7	..
1931	24.0	12.7	3.3	4.7	3.3	1.2	..
1932	24.2	12.8	3.2	4.6	3.6	1.2	..
1933	24.3	12.7	3.0	4.4	4.2	1.5	..
1934	23.1	12.7	2.0	4.2	5.5	1.9	0.5
1935	29.1	12.6	1.8	3.9	6.8	2.4	4.0
1936	34.7	12.3	1.7	3.8	8.9	2.9	8.0
1937	38.8	11.8	1.4	3.6	11.0	2.4	11.0
1938	44.4	11.3	1.3	3.5	14.3	2.3	14.0
1939	59.1	10.9 ‖	1.3	3.3	29.6	6.6	14.0
1940	77.3	11.4 ¶	1.2	3.3	47.4	22.0	14.0

* *Stat. Jahrb.* for various years; League of Nations, *Monthly Bulletin of Statistics* (1940), XXI, 304 and 408. Total Debt as well as Short-Term Debt includes tax certificates.
† Not included in Reich Debt; that is, includes only funds obtained on the "credit market" and not those from public sources, which represent merely a reallocation of funds among public bodies.
‡ Incurred before 1923, and revalorized bonds.
§ Estimated figures, probably lower limit. These estimates are compiled from various German sources: (a) *Der deutsche Volkswirt*, June 15, 1934, calculated by E. Wiedermuth of the Deutsche Bau- und Bodenbank; (b) *Frankfurter Zeitung*, December 29, 1935; (c) K. A. Herrmann's estimate in *Deutsche Wirtschaftszeitung*, April 28, 1938. The basis of these estimates is the increase in bill circulation.
‖ Estimated from *W. und S.* XIX (1939), 361 and 729.
¶ *D. V.* September 15, 1939.

of this sort, moreover, do not have the same significance as for a liberal economy. Such comparisons, however, indicate the conservative methods of financing which the Germans have used.

FINANCIAL POLICIES 159

Various sources of funds for armament expenditure are outlined in Table 15. Germany alone, excluding occupied and conquered territories, can probably maintain with satisfactory

TABLE 15

SOURCES OF FUNDS FOR GOVERNMENT EXPENDITURE
(Billion RM)

	National Income	Taxes and Duties of Reich, States and Municipalities *	Total Debt of all Government Units Including Secret Debt †	Savings from Decreased Expenditure on Unemployment ‡
March 31, 1929	76.0	13.5	14.1	..
1930	70.2	13.4	20.3	..
1931	57.5	11.9	24.0	..
1932	45.2	10.7	24.2	..
1933	46.5	10.6	24.3	..
1934	52.7	11.9	23.1	.4
1935	58.6	13.4	29.1	1.1
1936	64.9	15.5	34.7	1.5
1937	71.0	18.6	38.8	1.9
1938	79.7	19.1	44.4	2.4
1939	85 to 90 §	22.0	59.1	2.9
1940	100.0 ‖	27.0	77.3	2.9
1941	?	34.0	113.3 ¶	2.9

* *Stat. Jahrb.*, 1938, p. 635. Figures for 1939 compiled from *W. und S.*, 1939, pp. 149, 360, and 727. The figure for 1940 is an estimate contained in Deutsche Bank, *Wirtschaftliche Mitteilungen*, January 1940, p. 4, and the amount for 1941 is an estimate by E. W. Schmidt, director of the Deutsche Bank, *New York Times*, February 8, 1941.
† See Table 14.
‡ *W. und S.* for various years. Amounts in 1940 and 1941 are estimates based on the assumption that the rate of saving on unemployment expenditure has not increased.
§ *Frankfurter Zeitung*, December 10, 1939.
‖ *New York Times*, February 8, 1941.
¶ Estimate, *ibid.*

replacements a national income of 100 billion RM, thus maintaining indefinitely the present rate of expenditure of roughly 60 billion RM annually (see Table 16) — an amount derived by comparing depression income with the high level of income in 1940–41.

TABLE 16
Armament Expenditure *
(Billion RM)

	Taxes (Increase in each year over 1932)	Debt (Yearly increase)	Savings on Employment Insurance	Total
March 31, 1934	1.2	..	.4	1.6
" 1935	2.7	6.0	1.1	9.8
" 1936	4.8	5.6	1.5	11.9
" 1937	7.9	4.1	1.9	13.9
" 1938	8.4	5.6	2.4	16.4
" 1939	11.3	14.7	2.9	28.9
Pre-war total as of March 31, 1939	36.3	36.0	10.2	82.5
March 31, 1940	16.3	18.7	2.9	37.9
" 1941	23.3	36.0	2.9	62.2
War total	39.6	54.7	5.8	100.1
War and war preparation				182.6

* For sources, see Tables 14 and 15.

IX

REGIMENTATION AND CONSCRIPTION OF LABOR

IN SPITE OF the hatred and the attacks on the part of the industrialists and their allies against the system of collective bargaining and labor arbitration which had emerged as one of the chief results of the revolution of 1918, the democratic foundations of the labor constitution were preserved as long as political democracy existed. But after the democratic government was replaced by a dictatorship and independent labor unions were destroyed by that dictatorship, the time was ripe to build up the relationship between capital and labor on a new and rigidly autocratic basis.

On May 2, 1933, trade union centers were seized by Storm Troopers; union officials were jailed, and administration of the unions' property was entrusted to a government commissioner. Assurance was given to the workers that this action was directed only against their leaders, who were pictured as political traitors misusing workers' funds. In this manner the independent organizations of German labor, built up over several decades and enormously strengthened under the Social Democratic regime, were destroyed in one day.

The German Labor Front — After the process of "coördination" had been completed, the labor unions were reorganized into one large government organization, the German Labor Front. The National Socialist government recognized that destruction of the labor unions might strengthen radicalism among the workers and that it would be necessary to run these energies into channels useful to the dictatorship. A strong state organization could insure that employers' interests would not be sabotaged, and at the same time the social ties created by the unions could become valuable tools in the hands of the totalitarian state in diverting the "gaze of the masses from the material to the ideal values of the nation."

The legal basis for the German Labor Front was created in 1934 by the Law to Regulate National Labor, or, as it was hypocritically called, the Magna Charta of German labor. The organization was freed from all "class psychology" by opening membership to employers; members of free professions, business corporations, even villages and towns, trade associations, and the Reichs Food Estate are included in the Labor Front.

Membership is in principle voluntary but in reality compulsory. In any case all workmen, whether they pay dues to the Labor Front or not, are regarded as automatically members of the "works communities" in which they are employed. The "works communities" are factory units of the Labor Front organization, which is organized along regional, trade, and industrial lines. The employer of a business is called the leader, while his employees are followers. According to the "Magna Charta" the leader is master in his house and has the power to decide all matters concerning the enterprise. In this capacity the employer is told to "always use your capital in such a way as to make it yield profits corresponding to the national business." [1]

Politically the Labor Front is affiliated with and under the strict control of the National Socialist party and is regarded as the link between the party and all those of the German working population who are not party members. Dr. Robert Ley, a prominent member of the National Socialist party, is the leader of the Labor Front. All subordinate Labor Front officers are appointed by him and must hold membership in the party as an essential qualification of the appointment.

General Functions — Until the Bruening emergency decree,

[1] *News in Brief*, III, no. 11.

free collective bargaining was the normal way of settling wages. Arbitration by the government could only occur if all negotiations had failed. Strikes and lockouts were legal in so far as there was no collective contract, while illegal strikes and lockouts were not criminal offenses but merely subject to private damage claims. The emergency decrees allowed the government to determine wages. What started as an emergency measure the Nazis turned into a permanent principle. Above all they made all strikes illegal.

Another function of the Front is to "inculcate the National Socialist spirit into its members and to ensure contact between employers and employed on National Socialist principles." [2] It manages its own bank; provides vocational guidance with the end of raising the standard of efficiency of German workmen and the quality of work turned out, carries on its own welfare service, and participates in political parades and demonstrations. In order to imbue this vast mass organization with a new spirit, a new unit was attached, the "Strength through Joy" department. The administration of sport, recreation, and educational services for the working population was entrusted to it. Another branch of "Strength through Joy" activities is "Beauty and Labor," which works in coöperation with factory inspections for the eugenic and aesthetic improvement of working places. Even recreation after working hours is organized and supervised for the purpose of increasing the power of the party over labor. In February 1936 the department of "Evening Leisure" was set up to instruct the German worker how to spend his spare time in order to become a happy member of the National Socialist community.[3] No special fees are levied for the "Strength through Joy" organization, and Labor Front members are automatically members.

Mutual Trust Councils — In all firms with more than twenty

[2] Department of Overseas Trade, *op. cit.*, p. 204.
[3] Department of Overseas Trade, *op. cit.*, p. 210.

employees a "mutual trust council" composed of the employer-leader and from two to ten members of the factory unit is formed. The employer designates, after consulting with the National Socialist factory-cell organization, the candidates for this confidential council from among the employees. In 1934 and 1935 the workers were permitted to vote on these lists, but, if any of these candidates were defeated, the entire group was to be selected by the labor trustees, to be described below. Numerous such rejections brought suspension of election after 1935.

The council functions in a purely advisory capacity, acting as intermediary between employer and employed. It is expected to accomplish the following tasks:

1. Discuss measures directed towards increasing the efficiency of the firm.
2. Assist in the constitution and application of the works regulation provided by the employer.
3. Discuss the fixing of fines, the imposition of which are provided in the works regulations. Fines are usually imposed for the following types of offense: smoking on the working premises, uncleanliness, unpunctuality, waste of materials or electric power; and they may amount to as much as half the daily earnings of the offender.
4. Use their efforts to settle disputes regarding, for example, interpretation of the works regulation.

Labor Trustees — Labor questions in so far as is possible must be settled within the individual firm. Only when settlement within the firm has proved impossible are the disputes to be taken care of by arbitration machinery set up under the trustees of labor and by judicial process in the so-called "Social Honor Courts." Labor trustees are federal officials appointed by the Minister of Labor and are usually trusted party members. In each of the fourteen districts of the Nazi party there is a labor trustee who is authorized to perform the following functions:

1. Supervise the constitution and activities of Mutual Trust Councils.
2. Lay down general rules for work regulations. . . . The general rules are issued as recommendations, and employers are not compelled strictly to adhere to them but may take into consideration the conditions obtaining in their individual firms. The general rules have not the force of law, and no legal claim can be based upon them. On the other hand, an employer in drawing up the work regulations may only deviate from the general rules if he considers that their application in his business would be detrimental to it and to his employees. Willful departure from the general rules may lead to penalties decreed by the Social Court of Honor.
3. Issue wage schedules.
4. Approve of collective dismissals. If an employer wishes to dismiss over 10 per cent of his staff without reëmploying substitutes, he is bound to give notice in writing of his intention to the labor trustee of his district, who will examine the circumstances of the business in order to decide whether the dismissals are justified. Dismissal cannot take place until the labor trustee has given his approval.
5. Report to the Ministry of Labor on politico-social developments.[4]

Regulation of Working Conditions — In every firm employing at least twenty salaried and wage-earning employees work regulations and wage schedules must be issued in writing by the employer, who exercises legislative power delegated to him by law. Except as limited by general decrees and laws, and the wage schedules drawn up by the labor trustees, the employer has control over: (a) working hours; (b) wages and salaries; (c) schedules for calculating piece-work; (d) type, amount, and exaction of fines; (e) grounds on which employment can be terminated without notice.[5]

Courts of Social Honor — The practical importance of labor courts was greatly diminished by the abolition of collective bargaining and the destruction of the labor unions. But the

---
[4] Department of Overseas Trade, *op. cit.*, pp. 214–215.
[5] Department of Overseas Trade, *op. cit.*, p. 216.

special character of the semi-feudal relationship set up between "leader" and "following" required a special form of adjudication. Thus, fourteen "honor courts" were established to maintain discipline, and punish violators of the social honor.

Offenses against social honor are committed if:

1. An employer abuses his power in the firm, maliciously exploits the workers or "wounds their sense of honor."
2. A worker endangers industrial peace by maliciously provoking other followers.
3. A follower repeatedly makes frivolous or unfounded complaints or proposals to the labor trustee or obstinately disobeys written instructions.
4. Members of the Mutual Trust Council communicate without authority confidential information or manufacturing and business secrets.[6]

While employer and workers come under the jurisdiction of the Honor Courts, the primary emphasis is placed on offenses of labor against the employer and the establishment. Only the first rule pertains to offenses committed by employers, and there is a strong possibility that the employers proceeded against will represent establishments which have not enthusiastically sponsored the Nazi regime or owners of small shops whose influence would have little weight.

The Work Book — No clerical employee or manual worker may be hired who is not in the possession of a "work" book. The law covers practically all branches of employment, the whole field of large-scale industry, wholesale and retail trade and also banking institutions, and applies to apprentices and all recipients of wages or salaries less than 1,000 RM per month, totaling about twenty-one million people.[7]

Every worker is required to submit a work book if he wishes to continue in employment or to be employed (as from a date

[6] *Rgbl.*, 1934, I, 319.
[7] Department of Overseas Trade, *op. cit.*, p. 219.

that was specified for every trade, the first of which was for metal, building, engineering, and chemical industries as early as February 26, 1935). The book contains information regarding length and character of apprenticeship, vocational and other training leading to specialization, former positions held, present place of work, date when employment began, type of enterprise, and kind of occupation. In addition it tells whether the worker has a driving or flying license and what knowledge, if any, he has of agricultural work. There are two copies of this information: one in the hands of the worker and the other filed at the government employment office. The employment office thus has easily available information for a complete survey of the entire labor force.

The Labor Service — Two institutions of particular importance for National Socialist labor control are the Labor Service and the Land Help system. Although these institutions started as emergency measures to absorb the unemployed, both of them have taken on a permanent character. They are looked upon by the National Socialists as important tools in carrying out policies of promoting harmony among different social groups and strengthening the Back-to-the-Land Movement. Every young male German must serve for one year in the Labor Service before entering the army.[8] This year is one of preparation for the regular army service, and discipline in the Labor Service units is maintained according to military principles. Nazi proponents insist that this experience welds the youth of the nation into a real community, irrespective of class divisions, and inculcates the lesson of the dignity of manual labor. The men enlisted received a small allowance besides board and lodging. They are employed on various public projects, chiefly road and dam building, soil conservation, etc.

The Land Help system is not an organized unit such as the Labor Service, but is merely a way of transferring young

[8] *Rgbl.*, 1935, I, 769.

people, chiefly unemployed, from the cities to the country. The farmer who hires a city boy or girl as a farm helper receives a monthly financial allowance from the labor office. The labor office in turn saves an equal amount by paying relief to fewer persons. The land-helpers receive bed and board provided by the farmer and a small monthly cash benefit. Great efforts are made to keep these helpers on the farm longer than the usual six months' period. Marriages, "if racially sound," are encouraged with farmers' daughters and sons.

At the same time that the Hitler government decreed the reorganization of the labor market, a thorough transformation of social and relief policies was carried through. The legal provisions have not been changed fundamentally, but the dominating spirit has changed so completely that the unemployment insurance and relief system organized under the republic and the welfare policies of the dictatorship can hardly be compared. In the totalitarian state the bulk of relief is administered by a politically geared mechanism which is able to examine every individual case and to apply the strictest means test.

MOBILIZATION OF LABOR

Wholesale conscription of labor on June 22, 1938, almost one year and four months before the actual outbreak of war, reveals the extent of German experience with war-time labor measures. Moreover, the general law of 1938 was merely the end result of a system of motley partial conscriptions begun as early as 1934. Wage policy in war time was thus a logical continuation of peace-time policy. The chief problem which Germany faced was a shortage of labor in certain industrial areas and then, finally, a more general scarcity. To meet this problem, various measures were passed dealing with the distribution of workers among industries, vocational training and rehabilitation, and the mobilization of women, young people, the aged, Jews, and criminals.

The Employment Office — The government created a central agency to take over the tasks of alleviating labor shortages and directing the available supply toward industries and occupations engaged in military activity. This agency was already at hand in the Employment Service Board set up under the acts of 1922 and 1927. It covered the whole country through a central office and 300 district and local offices but did not have the exclusive monopoly of all hiring. Employment service was also performed by private nonprofit agencies, particularly by the trade unions. The Nazis, however, centralized all activities under the exclusive control of a government Employment Office. By act of November 5, 1935, the Employment Office was granted full monopoly of employment service, vocational guidance, and the placing of apprentices, leaving to private agencies only the employment service for actors and musicians. All non-profit agencies ceased functioning on July 31, 1936.[9]

Labor Feudalism — By gradual steps, government decrees which provided for mobilization of workers in industries and occupations engaged in military work created a modern equivalent to medieval feudalism. The serf of the Middle Ages was considered part of the estate of his squire or lord. He was attached and fixed to the estate and had no right to move away. The German worker also has become attached and fixed to his job, whether agricultural or industrial. The successive stages in this development will next be described.

According to a decree of May 1934, the agricultural worker, although not as yet fixed to a particular estate, was nonetheless fixed to the agricultural occupation. Industrial employers under this decree were ordered to discharge workers who had within the preceding three years been engaged in agricultural work. In addition the legislation attempted to prevent a continuation of the exodus from the country that was still continuing. No person engaged in agriculture within a specified period

[9] *Rgbl.*, 1935, I, 1291.

was permitted to be employed on a non-agricultural job except by special permission of the government Employment Office.[10]

No sooner had measures been taken to alleviate the shortage of agricultural workers than scarcities also appeared in a different sector. Manufacturing of guns, tanks, and warships had progressed on such a scale by the end of 1934 that a scarcity of metal workers appeared. And on December 29, 1934, all skilled metal workers were required to obtain consent of their resident Employment Office before obtaining work outside the district.[11] The skilled metal worker was thus fixed to the district of his Employment Office, tied to one particular geographical area. Similar decrees extended the scope of this measure as follows:

November 27, 1936 — Unskilled metal workers [12]
October 6, 1937 — Carpenters and masons [13]
May 30, 1938 — All building workers, including construction engineers [14]
March 10, 1939 — Forestry, mining, chemical industry [15]

The circle of persons regulated by the decree of March 10, 1939, includes family relatives employed non-contractually by husband or wife, father or mother, grandfather or grandmother, brother or sister. They are required as of May 1, 1939, to keep a work book like any regular wage earner. Apparently the main purpose of including unpaid workers was to prevent relatives of farmers from giving up unpaid work in the country for a job in the city.

The worker under the Nazi system has thus become the industrial equivalent of the medieval bondsman. The feudal lord could demand payment to recompense him for a serf who had broken bond by leaving the ground he was attached to; similarly, the Nazis have made it a criminal offense for the worker

[10] *Rgbl.*, 1934, I, 127–128.
[11] *Rgbl.*, 1934, I, 202.
[12] *Rgbl.*, 1936, I, 311.
[13] *Rgbl.*, 1937, I, 248.
[14] *Rgbl.*, 1938, I, 191.
[15] *Rgbl.*, 1939, I, 444.

to quit his job without permission of the employment authorities. Sentences ranging from two to eight months of imprisonment are common.

Mobilization of Independent Entrepreneurs, Women, the Aged, Jews, and Criminals — From the outset, the Nazis proclaimed hostility to nondomestic activities of women, but in this field, as in so many others, they chose to reverse their policy. This reversal, like other changes of policy, was carried out unobtrusively and inconspicuously. The first step was to take away social insurance benefits from wives of soldiers if they were available for employment but refused work.[16] In October 1937 women who received marriage loans were no longer forbidden employment in industry.[17] But not until February 15, 1938, was provision made for systematic utilization of the reserve of women workers. At that time a year of compulsory labor service for women was introduced. Unmarried women under the age of twenty-five years could be engaged in the clothing, textile, and tobacco industries, or employed as commercial or office workers, only if they had been employed at least one year in agriculture or in domestic service or had been engaged for two years in nursing, welfare service, or kindergarten work, as shown by their employment books.[18] As time went on, the supply of hands, especially in agriculture, became such a problem that on December 23, 1938, every young woman was required to serve a year in the compulsory service before entering any trade whatsoever.[19] In addition, employment of women was subject to consent by the employment authorities.

Although these measures drew a number of women into employment, the number was not large, and in the spring of 1939 additional measures were announced for "planning allocation of female labor." Each German woman of working age, mar-

[16] *Rab.*, 1936, I, 327.
[17] *Rgbl.*, 1937, I, 1158.
[18] *Rab.*, 1938, I, 46.
[19] *Rab.*, 1939, I, 48.

ried or unmarried, was required to fill out a questionnaire giving full details of her abilities and capacities. On the basis of these questionnaires, the Nazis started a drive to force women to enter employment. The success of this drive was announced by the Minister of Labor on August 15, 1939.[20] He stated that in two years' time the number of women gainfully employed had increased by 18 per cent as compared to 10 per cent for men. This meant that 32½ per cent of the total gainfully employed were women.

In addition to women, the Nazis mobilized the aged. On January 1, 1939, those who reached retirement age in relative health could no longer claim their benefits under social insurance, and workers already retired were forced back to employment. Special workshops or shop sections were set aside to utilize their reduced speed and efficiency.[21]

As a result of the diligent search for additional hands, employment opportunity was at last opened for the Jews, who had for years suffered severe discrimination. A circular issued by the Employment Office during February 1939 urged business to engage Jewish workers as quickly as possible in order "to release German (*sic*) workers for urgent construction work," and decreed that contractors and undertakings would no longer be penalized because they employed Jews.[22] The number of Jews who have thus been employed is, however, small, and the new decree in no way indicated that the fate of the German Jews had been changed, for it was clearly specified that Jews should work separately from the rest of the workers "in order to minimize contamination."

Convicts were not overlooked in the general drive to increase the labor force. On May 18, 1938, the president of the Employment Offices decreed the introduction of compulsory labor

---

[20] L. Hamburger, "How Nazi Germany Has Mobilized and Controlled Labor" (pamphlet, Brookings Institution, 1940), p. 35.
[21] *Ibid.*, p. 37.   [22] *Arbeitsrecht*, March 10, 1939.

for all prisoners regardless of length of term or character of crime. Provision was made for retraining convicts who had previously been employed in metal trades. Convicts were to be employed in quarries, brick factories, mining, cable-laying, forestry, power plants, soil-conservation projects, road building, and in the canning industry. In view of the fact that convict labor would come in contact with free labor, the regulations stated that convicts should be grouped by teams of ten, except under special circumstances when employment of convicts individually, without supervision, is permitted. The employer provides for maintenance not only of the worker but also of the guards, and pays the government 60 per cent of the normal wage — payment for upkeep is counted as part of the 60 per cent. The employer thus gets cheap labor, while the government has an excellent arrangement for avoiding prison costs, even earning something from the work of the convicts.[23]

Upon the adoption of an open armament program, handicraft enterprises were discouraged as all the available supply of labor was needed in large-scale industries. A retroactive law of January 1935 thus provided that only a duly registered master artisan could own an independent enterprise — all independents not masters had to pass a master's examination or seek employment in industry.[24] This decree was buttressed by more far-reaching provisions on February 22, 1939,[25] which forbade 15,000 craftsmen from continuing in handicraft trades. Those engaged in trades considered by the Nazis to be overcrowded, e.g., bakers, butchers, hairdressers, tailors, and shoemakers, were no longer licensed to do business. A craftsman who was thus struck off the handicrafts register could apply for fresh registration only after three years. In no case was an indemnity given for any losses.

[23] *Rab.*, 1938, I, 207.
[24] *Monthly Labor Review*, April 1938, p. 888.
[25] *International Labor Review*, April 1940, p. 305.

Other laws were aimed at directing the million itinerant workers into the armament industries. The Itinerant Trades Taxation Act of December 10, 1937,[26] increased the taxes on hucksters and drove some of them into industrial employment. Finally a decree of January 29, 1938, provided that permits to engage in itinerant occupations or in hawking might be refused or withdrawn for persons whose labor could be more usefully employed elsewhere.[27]

Apprenticeship and Training — The training of young workers has largely been left to private enterprise, while the retraining program has been carried out by the government Employment Office assisted by the Nazi party and its affiliated organizations — the Labor Front, the National Socialist Welfare Organization and the Reich Food Estate. All training courses are auxiliary measures in the systematic organization of the employment market, and thus even private training programs were subjected to government regulation.

By a decree of November 7, 1936,[28] all metal and building firms employing ten or more men were required to train a number of apprentices bearing a "reasonable" ratio to the number of skilled workers in the firm — the Employment Office to determine what was a "reasonable" ratio. Employers who found it difficult or impossible to employ the number of apprentices were forced to pay a commutation fee amounting to 50 RM for each missing apprentice. For all other industries, the regional Employment Offices were authorized to compel firms to maintain the same percentage of apprentices as the average for the industry in question.[29]

Government subsidies to private enterprise are used in the retraining program. For example, the government will pay the

[26] *Rgbl.*, 1937, I, 1347.
[27] *Deutscher Reichs- und Preussischer Staatsanzeiger*, no. 25, 1938.
[28] *Ibid.*, no. 202, 1936.
[29] Helmut Vollweiller, "The Mobilization of Labour Reserves in Germany: II," *International Labor Review*, XXXVIII (November 1938), 596.

worker an unemployment benefit in addition to a learner's wage received from the employer. The employer decides within six weeks whether he wishes to keep the worker for not less than five months for additional training and employment; if he does, the worker receives the unemployment benefit for a period of not more than eight weeks. If the worker cannot be trained in the enterprise itself, arrangements are made for ordinary courses in established schools or in special residential courses — agricultural or domestic training camps, auxiliary trade and instruction camps. Attendance at such courses is compulsory for certain groups of workers who receive unemployment benefits. Especially important work is carried on by the Labor Front, which organizes courses of practical work, training workshops, and training camps where the semi-skilled and the unskilled retrain or obtain the necessary knowledge for a new occupation.

These measures are not designed to increase the supply of highly skilled workers who require a vocational training of several years. The government program is one merely of increasing the supply of semi-skilled workers in occupations where there is marked labor shortage and in which there are immediate vacancies for employment. But the government hopes in this way to release skilled workers from tasks where their skills are partly wasted because energies and abilities are being used at semi-skilled work.

It should be noted that the program of training began as early as 1936 and thus bore fruit by the time war broke out. In the beginning the Employment Offices had no difficulty in supplying apprentices on a voluntary basis, but by 1938 it was necessary to use compulsion. According to a decree of March 1, 1938, all young people wishing to enter occupations as apprentices were required to get permission from the Employment Office.[30] The Office could thus direct recruits toward trades

[30] *Rab.*, 1938, I, 69.

and occupations essential for war preparations. Even young people seeking work without asking for wages or salaries were included. Parents or guardians were ordered to report to the Employment Office all young people leaving primary or secondary school. For all practical purposes control over new entrants into industry had become complete even by the spring of 1939. The war regulations of September 1, 1939, thus included no new provisions when they required all apprenticeship to be subject to authorization by the Employment Office.

Labor Conscription — The various measures dealt with in the preceding discussion failed to reach every labor reserve, and the Nazis resorted to conscription on June 22, 1938.[31] Under this decree, conscription was extended to all Germans of every age and type, "whether man or woman, schoolboy or aged, employer or worker, civil servant or business man." The president of the Employment Offices, as chief recruiting agent, was authorized to conscript workers for any place he chose for a specified period not to exceed six months, or to require any worker to undergo a specified course of vocational training. The general regulations covering employment and social insurance also applied to the new service or training.

A new order of February 13 and March 2, 1939, extended the scope of conscription — aliens were included, and the period of service might be extended for an indefinite period.[32] This order represents in truth an order for civil mobilization given six months in advance of the beginning of hostilities.

Sweeping decrees issued in September 1939 forbade all workers and employees to leave their jobs without the consent of the local Employment Office, lifted all restrictions on hours of work for adult men, abolished restrictions on night work for women, and removed limitations governing the employment of women and children under eighteen.[33] These stringent

[31] *Rgbl.*, 1939, I, 652.
[32] *Rab.*, 1939, I, 126.
[33] *Rgbl.*, 1939, I, 1685, 1690.

decrees, however, brought increased accidents, caused discontent, and fatigue which slowed up the tempo of industrial production. Therefore, a new decree of December 12, 1939, limited the working day to ten hours and with special permission, to twelve. Night work for women and children was again prohibited, and their working week restricted to fifty-six hours.[34]

In order to utilize all available labor to the fullest extent, employers were ordered to report any skilled workers whom they could release.[35] All business firms were compelled, on demand of the Employment Office, to hire workers for retraining and to employ women for work formerly done by men. Commissions were named to study the most effective use of workers.[36] Special efforts were made to mobilize housewives for part-time work. Children leaving school were conscripted into occupations in which scarcities were most severe. The most serious problem facing the government with the outbreak of war was the lack of trained engineers and supervisory forces, and these deficiencies could not be easily repaired.

[34] Deutsche Bank, *Wirtschaftliche Mitteilungen*, December 1939, p. 271.
[35] Deutsche Bank, *op. cit.*, November 1939, p. 255.
[36] *D.V.*, February 9, 1940.

X

AGRICULTURAL "PLANNING"

WITHIN two months after his accession to power, Hitler announced to his party comrades that the creation of a peasant state of "race and soil" was essential as a bulwark against Bolshevism: "There is only one final last chance for the German peasantry! Following this administration, logically only one other can come, Bolshevism! . . . But if this regime can carry through the objectives which I have laid before you, then the peasantry will become the supporting foundation for a new Kingdom of Race and Soil." [1]

The program contrived by the Nazi government for the purpose of establishing this peasant kingdom has three basic elements: fixity of occupation, of status, and of residence. The peasant has been bound to the soil in a rigid and permanent relationship, and fitted into a class hierarchy, unalterable, except at the discretion of the state. Both immigration and emigration across national borders have been prohibited except with government permission, while movements to and from cities and between agricultural areas have been rigidly controlled. These elements of German agriculture rigidly controlled, the Nazi leaders hoped to achieve two major objectives — "nourish the race with men" and supply the people with food.

The following analysis of the Nazi Farm Program will concentrate on the three most important aspects: (1) the Hereditary Farm Act, under which peasants with German citizenship and pure Germanic blood are attached to the soil rigidly and

---

[1] Quoted in Brady, *op. cit.*, p. 213.

forever; (2) the Reich Food Estate Act, which centralizes and controls administration of all phases of agricultural production and distribution; (3) the Rural Resettlement program, which aims at keeping the agricultural population fixed on the land and decentralizing the industrial population of the large cities. None of these measures was without precedent. For example, the emergency measures of pre-Nazi government protected the farm against foreclosures and granted subsidies and extraordinary credits to agriculture, especially in the eastern districts of the large Junker estates. Agricultural products had been protected by high tariff walls; attempts had been made to raise prices by fixing compulsory quotas for certain goods through milling provisions.

One of the important factors leading German Republican forces to defeat after fourteen years of struggle was their missed opportunities in the land problem. The political power of the Junkers was undeniable and represented the continuation of a feudal caste system. Six per cent of all the landowners in Germany owned 24 per cent of the land; the 412 largest landowners owned as much as 1,000,000 peasant proprietors.[2]

The Social Democrats promised to subjugate this seigneurial caste by raising the wages of farm labor, increasing taxes, and lowering tariffs on grain. The universal and equal franchise for men and women, the abolition of the large estates as units of local government, and the abolition of the entail also were proposed in order to serve the same purpose, especially in combination with governmental purchases of large estates and the division of them into small holdings. The interconnection between political solutions and economic forces is clearly evident here. Nothing reveals the truth of this better than the transition from the Social Democratic act empowering the govern-

[2] Announcement of Minister of Agriculture, Herr Darre, requesting the Reich Statistical Office to make a survey of the distribution of land: *London Times*, September 2, 1936.

ment to confiscate farms and land for the purpose of settling small farmers to the policy of ultra-protection and agrarian subsidies initiated by the same Social Democrats. They had finally succumbed to the pressure of the politically strong Junkers. Thus, of the two billion RM spent as direct or indirect subsidies to agriculture between 1926 and 1930, the lion's share went to the eastern estates. Above all, the entail remained.

In 1928 one-third of the large estates were bankrupt, but, when Chancellor Bruening in 1932 suggested their liquidation, Hindenburg dismissed him with the statement that he did not want any agrarian Bolshevik experiments.[3]

THE HEREDITARY FARM ACT

To win the support of the small peasants, fascism made demagogic demands for the division of the big estates. But the first of the three important measures carried out by the Nazis in agriculture, the Reich Hereditary Farm Act of September 29, 1933,[4] had the effect of protecting large and medium-sized farms at the expense of the small peasantry. This protection was to be expected in view of the fact that large landholders had given support to the Nazis in their final push to power.

Moreover, the first Minister of Agriculture under Hitler was the representative of the big landowners, Hugenberg, and his rival and successor, Walter Darré, although professing to be a defender of the small peasant, followed his predecessor's policy. "In agreement with the Chancellor," Darré declared, "I shall not touch any property, whatever its size, if it is economically sound and can support itself."[5] Hitler had stated emphatically a month or so earlier that "large rural property has the right to

---

[3] Karl Brandt, "Junkers to the Fore Again," *Foreign Affairs*, October 1935.
[4] *Rgbl.*, 1933, I, 685, 749, 1096.
[5] *Le Temps*, July 31, 1933.

exist legally on condition that it is worked for the common good of all citizens."[6] He appointed as commissioner for resettlement another representative of the Junker landowners, Baron von Gayl, a former minister of the Papen government.[7]

The basic ideas of the Reich Hereditary Farm Act are the following: Farms large enough to be self-sustaining, that is, to support one family, are called hereditary estates if they are the property of a person who is legally entitled to be a peasant. German citizenship, pure Germanic blood since January 1, 1800, and an honorable character are required to obtain this legal status. All farms of the prescribed size and owned by a peasant automatically became hereditary farms. Although it was intended that these farms should not in general be larger than 125 hectares, large estates can also be registered as hereditary estates upon special application, if there is a public interest involved. In view of the fact that an hereditary estate cannot be sold, mortgaged, or foreclosed, numerous owners of large estates succeeded in registering as hereditary estates in order to be free of any interference from their creditors.[8]

By January 1, 1935, about 700,000 farms out of about 5½ million in Germany had been registered as hereditary estates. They were thus proclaimed inalienable and could be inherited only by a single heir (the oldest or the youngest son, according to the region) — a provision that prevents division of the property.

In order to establish hereditary holdings sufficiently large to meet the requirements of the act, the Nazis in many districts confiscated small farms or took the use of certain lands from the poor peasants. For example, the government of Baden by a decree of February 1934 withdrew from the peasants their age-old right to use for pasturage the communal lands repre-

[6] *Journal du Commerce,* June 29, 1933.
[7] David Guerin, *Fascism and Big Business* (New York, 1939), p. 265.
[8] Guerin, *op. cit.,* p. 265.

senting 17 per cent of the area of Baden in order to create "hereditary farms." A government decree of December 27, 1934, in Hesse, expropriated in the same way and for the same

TABLE 17

CONCENTRATION OF OWNERSHIP IN AGRICULTURE
(*Percentage of total number of farms*)

Size of Farm (hectares)	1933	1938
Under 15	50.6	43.1
15 to 50	43.6	49.5
50 to 75	3.9	4.8
75 to 100	1.3	1.7
Over 100	.6	.9

SOURCE: *Vjh. Stat. d. D. R.*, 1939, III, p. 8.

purpose 192,000 hectares of peasant lands (13.8 per cent of the area of that state). "In the swampy region of the Rohn a drainage plan was adopted, the sole purpose of which was to expropriate tens of thousands of wretched peasants with tiny holdings in order to put a few hundred 'hereditary farmers' on the improved land." [9]

TABLE 18

NEW PEASANT HOLDINGS ACCORDING TO SIZE, 1919–38

Year	Total Number	Less than 2 hectares	2 to 5 hectares	5 to 10 hectares	10 to 20 hectares	20 and over
		PERCENTAGE OF TOTAL NUMBER				
1933/38	3401	5.1	6.7	15.3	51.6	21.3
1919/32	4104	29.4	11.2	14.1	36.8	8.5

SOURCE: *Vjh. Stat. d. D. R.*, 1939, III, 6.

The effect of the Hereditary Estate Act is summarized in Table 17, which shows a striking growth in the share of large and medium-sized farms in the total number of farms.

According to statistics given in Table 18, the number of new peasant holdings, moreover, showed a remarkable decline in the

[9] Guerin, *op. cit.*, pp. 265–266.

percentage of the total number in very small farms and an equally notable increase for holdings above 10 hectares. Whereas 8.5 per cent of the total number of new peasant holdings were 20 hectares or over for 1919–32, holdings of this size came to 21.3 per cent of the total number for a six-year period under the Nazis.

The Central Administrative Division I of the Reich Food Estate enforces the inheritance law; all disputes regarding the rights of inheritance are decided by the Estate Courts, which are subject in the final analysis to the administrative control of the National Peasant Leader.

THE REICH FOOD ESTATE

During the course of 1933 and 1934 agriculture was reorganized under the Reich Food Estate, which is a self-administrative statutory corporation comprising all individuals and organizations concerned in the distribution as well as the production of agricultural commodities.[10] It covers not only farming proper but also forestry, market gardening, fisheries, game questions, and viticulture. On the distribution side, agricultural coöperative societies and all processors and traders in agricultural products, whether individuals, firms, or associations, are included. Thus millers, brewers, bakers, confectioners, and butchers belong to it,[11] as well as to the Estate of Handicrafts or Industry and Trade. The Food Estate is competent for all matters affecting production, sales, and prices of their products.

All previously existing organizations, such as the chambers of agriculture, the National Farmers' Association, agricultural coöperative societies, and so on, have been either dissolved or

[10] Act authorizing the preliminary establishment of the Reich's Food Estate and measures for the market and price regulation of agricultural products, September 13, 1933 (*Rgbl.*, I, 626). By an act of July 15, 1933 (*Rgbl.*, I, 495), jurisdiction over matters of agricultural organization was transferred from the states to the Reich.

[11] *Rgbl.*, 1934, I, 100, 259, 527.

incorporated into the Food Estate. At the same time, marketing associations (*Marktverbaende*) have been set up and along with these government boards (*Reichstellen*) which are responsible for regulating farmers' and processors' prices as well as the import and export of principal foodstuffs. The top-ranking officer of the Food Estate is the "farm leader," who actually, but not of constitutional necessity, is also the Reich Minister for Food and Agriculture. The Ministry for Food and Agriculture exercises the state administrative functions which fall to it by law, while the Estate "is conceived rather as the co-ordinated self-expression of the members of the agricultural and commercial bodies and individuals closely concerned with agricultural products." [12] Under the leadership principle, the farm leader is head of all departments, officials, employees, and members of the Estate; his authority and responsibility are supreme. Nominations are made from the top down; responsibility is from the bottom up.

The market association separated according to main agricultural products are in turn organized on regional lines. Regional divisions are drawn together in a central association which directs and supervises production and distribution from the earliest stages of production until the goods reach the consumer. Horizontal federations including all producers of the same stage in the productive process are empowered to license the creation of new undertakings or the enlargement of existing ones. In addition, they enforce trading and marketing in accordance with standards of quality, determine profit margins for the wholesaler, manufacturer, and retailer, and exercise general control over distribution.

The government boards regulate producer prices and are different from the market associations in being trading bodies which may be entrusted with the actual purchase and sale of

---

[12] Department of Overseas Trade, *op. cit.*, p. 40.

AGRICULTURAL "PLANNING"

both domestic and foreign produce.[13] A permit is required before any imported food is sold within Germany. In exchange for this permit, the importer must give the board the difference between the higher internal price and the price of foreign foods, including duty and transportation costs. Domestic producers must also obtain a permit before selling domestic goods. Moreover, they may be required to supply foodstuffs to the board at prices fixed by the latter. Controlling as they do the amounts of foreign foodstuffs and the prices at which they are sold in Germany, as well as the accumulation of domestic foodstuff surpluses, government boards thus exercise a determining influence upon internal food prices.

The organization of the dairy industry as a typical example of market regulations will be described briefly.[14] The producers and dealers in dairy products are organized in a central marketing association with seventeen regional milk supply groups. Obligations of the milk producers to produce certain amounts of milk, to deliver them to a certain dealer at a certain date and at a certain price, are defined by the directors of the regional associations assisted by advisory councils. A violator of an order of the association is liable to punishment by fines, and the order can be enforced through removal of the permit. A government board for milk, oil, and fat has been established with seventeen regional subdivisions. Its chief task is to balance the surplus of agricultural regions with the demands of industrial cities and to provide for means of transportation. It has a monopoly on the importation of dairy products of foreign origin for distribution in Germany. A Reich commissioner, appointed by the Minister for Food and Agriculture, super-

---

[13] Benjamin Higgins, "Germany's Bid for Agricultural Self-Sufficiency," *Journal of Farm Economics*, XXI (May 1939), 435–446, discusses price and production control in greater detail with reference to grain, livestock, poultry, and dairy products.

[14] *Rgbl.*, 1934, I, 259.

vises the entire national market for dairy products.[15] Important rules and regulations issued by any of the associations require the approval of the commissioner. The commissioner also determines the ratio of agricultural raw materials to be used in certain goods, for instance, the percentage of milk in chocolate and ice cream, or of fat in artificial butter. Ordinarily, no peasant is allowed to sell these goods directly to the consumer.

During the first year of National Socialism, the government endeavored to protect the farmer against falling food prices and even to raise the price of agricultural commodities, particularly in grains. Although it has been from the very beginning a cardinal feature of Nazi policy to keep the price level as a whole fairly stable, one of the tasks of the Reich Food Estate was to close the gap between the price levels of agricultural and industrial products. This task was considered essential not only from an economic standpoint, but also for social and political reasons.

The years of depression since 1929, which bore particularly heavily on agriculture, did not affect all farm commodities equally. Pre-Hitler protective tariffs favored various grains, especially wheat and hence the large farmers of the east, at the expense of animal products. During the period between 1929 and 1933, agricultural wholesale prices decreased by 33 per cent, with the price of grains falling only 21.8 per cent, whereas the price of slaughtered animals declined by 50 per cent.

In 1933–34 the Hitler government intervened to prevent the large harvest from depressing grain prices. Millers were forced to buy and store grain to an amount equal to 150 to 200 per cent of the quantity milled by them for the monthly average of 1932–33. In addition, the government purchased and stored a portion of the crop.[16] Although other farm prices were main-

[15] *Rgbl.*, 1934, I, 198.
[16] These accumulated stocks, however, were drawn upon in 1934–35, a

tained practically constant at the 1935 level, cereal prices were increased further as a result of a series of medium harvests, together with attempts to reduce imports.

For agricultural prices in general, it can be said that, whereas in 1933 the producer was usually guaranteed a minimum price, later the producer could not charge more than a fixed maximum price. The scarcity of foreign currency and curtailment of food imports brought with it a scarcity of foodstuffs such as fat, eggs, and butter and made control urgent. In spite of the comprehensiveness of the agricultural administration, the government had difficulties in keeping food prices at the desired level, and in 1934, when a price commissioner was appointed to fix prices for all commodities and services, the Food Estate was also put within his jurisdiction.

Another method of food and price control which has been used widely since 1936 is the so-called steering of consumption (*Verbrauchslenkung*). Propaganda instructs consumers to use more of one kind of foodstuffs in order to consume less of another commodity which is scarce. German authorities advise the public to decrease its consumption of butter and other fats and use more jam. Local shop displays, special recipes in the press, the program of the cookery schools of the Reich, the associations and affiliated bodies of the Party, such as the National Sozialistische Frauenschaft and Deutsches Frauenwerk, have been drafted to carry out the propaganda. Rations of the Labor Service reflect the surplus-scarcity situation, and fish days have been introduced in colleges and other institutions.

It is extremely difficult to find out how the vast and complicated organization of the Reich Food Estate functions in detail. Theoretically, the control is complete over all phases of agricultural life — social, cultural, technical, and economic.[17] But

---

year with a poor cereal harvest. Hence for 1934 there was a larger proportional increase in prices of animals and animal products than for cereals.

[17] The most important tasks of the Food Estate were: the integration

how far changes have been made in the actual operation of agricultural activity is another matter. Some observers maintain that agriculture has been transformed from a sector of an individualistic capitalist society into a semifeudal order.[18] Guillebaud, on the other hand, suggests, along with the German official point of view, that the form adopted is that of "self government under the supervision of the State in the interest of the community as a whole." [19] It has been further suggested that the idea behind agricultural control is similar to the AAA in the United States.[20]

Fixity of occupation, status, and residence fulfill functions that in the feudal society were accomplished by the ties of feudal loyalty given by the masses to their seigneurs. The Reich Food Estate establishes close connection between the central authority and the individual peasants. The claims for "self-government" in the Estate are put forth on the grounds that few changes were made in the actual structure and functioning of the already existing associations and federations. Although they were simply taken over intact, membership was made compulsory instead of optional and the leadership principle was substituted for the elective rule. The most important result of the new set-up was to increase enormously the centralization and coördination of control — a development which in pre-Nazi days had gone further in industry than in agriculture. The Estate dealing chiefly with production and distribution of goods, with quotas and commodity prices, constitutes a highly rationalized organization with important features similar to those of an industrial cartel.

---

of the German peasantry into an organic unit, the regulation of the economic and social affairs of its members, the reconciliation of different interest groups in agriculture, the education of the peasant in order to make him conscious of his "mission" in the nation's life, and coöperation in the administering of the Hereditary Estate Act (*Rgbl.*, 1933, I, 626).

[18] Fritz Ermarth, *The New Germany* (Washington, D. C., 1936), p. 97.
[19] Guillebaud, *op. cit.*, p. 58.   [20] Higgins, *loc. cit.*, p. 437.

One of the goals of the price control was to narrow the gap between agricultural prices and industrial prices in order to guarantee profits in agriculture and in this respect resembles the Agricultural Adjustment Administration. The extent of state control was considerably more far-reaching in the German case. A program of agricultural self-sufficiency favored the large landholders who had subsidized Hitler in his rise to power, for the big estates growing grains profited most from the price margins given to that product and lent themselves more easily to intensive, scientific, mechanized farming than small tracts of land. Moreover, the creation of medium-sized hereditary estates for the purpose of securing a dependable social base for the Nazi party was done at the expense of the small landowner rather than the large one, who actually used the hereditary act to protect his large estate against creditors.

RURAL AND SUBURBAN SETTLEMENTS

The Social Democrats made constant efforts to settle peasants on small and middle-sized farms, but had meager success because of the many political difficulties — the iron resistance of the Junkers and the objection of left-wing Social Democrats who did not want to see the peasant element strengthened.[21] The Nazis also proclaimed rural resettlement as one of their main goals, but later realized that rural resettlement was a slow and expensive procedure. Hence they focused attention on plans for suburban resettlement as a solution for problems connected with population and race, labor unrest, and agricultural self-sufficiency. Suburban settlements are of two general designs: those grouped around large industrial plants and settled primarily by company employees, and the newer type introduced by the government during the depression, those

[21] Karl Brandt, "The German Back-to-the-Land Movement," *Journal of Land and Public Utility Economics*, May 1935.

scattered on the edges of the larger industrial cities.[22] Governmental residential settlements, however, closely resemble company mill-town arrangements.

Even prior to the Nazis, there were a good many mill towns similar to Gary, Indiana, and Longview, Washington. These towns varied from the completely self-contained villages of the Krupp and Siemens plants to the supplementary housing commonly seen in the Rhineland coal fields. The company sold to a factory worker a plot of ground large enough for a house and garden, and the transaction was financed through the company, participating commercial companies, or through some government agency. The Krupp plan, regarded by many as ideal, provided that necessary loans must be secured by insurance policies taken out by every prospective settler and made over to the company. Deductions from payrolls guaranteed the payments of premiums and of interest and amortization on loans.

Extension of company programs was urged by the Nazis. In the Rhenish Westphalia district, for example, at the end of 1937, industrial enterprises controlled 240,000 houses. Thirty per cent of the funds needed for building the houses was furnished by the employers. At the Essen works alone, Krupp had constructed 29,000 houses by October 1937 and had also bought stock in three building coöperative associations and one building corporation.[23]

The government's encouragement took the form of outright subsidies, loans at a low rate of interest, and tax exemptions. In order to obtain this assistance, the cost of the house, exclusive of land, might not amount to more than 7,000 RM, of which 2,000 RM could be obtained from a government loan carrying a rate of interest of 3 per cent. According to the Real Estate Law of 1936, homes bearing a monthly rental of 40 RM

[22] *Rgbl.*, 1933, I, 659, and *Rgbl.*, 1934, I, 568.
[23] *Monthly Labor Review*, January 1939, pp. 99–102.

or less and completed between April 1, 1937, and March 31, 1940, were made tax exempt for twenty years.

The amounts involved in the government suburban settlement program were small. From late summer 1935, when the program was inaugurated, until September 1938, a total of only 446 million RM had been spent in this way, while the total area of land utilized came to 9,081 hectares, or 0.032 per cent of the total area under cultivation.[24]

In support of the suburban settlement program, business claimed the program would ameliorate labor unrest by giving labor a property interest, cut down on costly factory turnover, and make lower wages possible. The settler, being indebted to the plant or finance and insurance companies whose claims are guaranteed by the industry founding the community, is practically attached to the plant. Subsistence in case of depression and factory lay-offs is guaranteed by the plot of ground. Moreover, company towns provide business for private loan, insurance, and building-supplies companies which may or may not be subsidiaries of the company promoting the plan. The Nazi party in turn found certain advantages in the plan for themselves: the neutralization of labor by giving it something to do during off-work hours.

A further measure to reëstablish the "broken ties between men and the soil" was the land-helper system administered and financed by the labor offices. At the same time it enabled landowners to get cheap labor. Young men and women from the cities were placed by the government employment offices on farms in return for the food and shelter which the farmer provided and a small sum of pocket money from the employment office. By the decree of August 28, 1934, unmarried men under twenty-five lost their jobs in the cities and were sent to the country as agricultural helpers. The state also placed at the disposal of farmers the members of the Labor Service, who,

[24] *W. und S.*, XIX (1939), 734.

by the law of April 1, 1934, had to spend a year on the land after leaving school.

The conditions of the rural workers apparently were so bad that, even when they had work, they left their villages and poured into the cities. The law of May 15, 1934, strictly prohibited urban businesses from hiring employees who had worked in agriculture during the preceding three years, and a decree of February 28, 1935, provided that farm workers to whom the previous law applied should be expelled from the cities immediately and sent back to the country on pain of criminal prosecution.

Tax exemptions, subsidies, and debt relief measures as applied to agriculture were aimed more at the general problem of depression than at the more long-run one of establishing a kingdom of "race and soil." These measures, moreover, tended to favor the well-to-do peasants rather than the little dirt farmer. For example, the hereditary farms were wholly exempt from the inheritance tax and the real estate tax. The larger farms also profited from tax exemptions for the acquisition of new machinery, automobiles, construction of new buildings, and the like. The working peasants on the other hand were burdened more heavily by the poll taxes newly introduced in 11,000 communes.

To the Junkers and rich peasants, the Nazi government continued to distribute the eastern provincial emergency subsidies (*Osthilfe*). Before Hitler took power, the Nazis had denounced the fact that out of 132 million RM distributed in December 1932, 60 millions had gone to farms of more than 100 hectares. Once in power, however, they pursued the same policy: by November 1934, 213 million RM had been paid to farms of more than 125 hectares, 194 million RM to farms between 7.5 and 125 hectares, and only 33.5 million RM to farms of less than 7.5 hectares. Furthermore, the subsidies to increase agricultural production amounting to a total appropriation of over one billion RM go chiefly to the large and medium-sized landowners, who are in a position to expand or undertake intensive production of the desired crops.[25]

Although the importance of an adequate organization for making credits available is recognized, progress in this direction is retarded by the fact that existing indebtedness has not all been properly assessed nor have provisions been made for amortization. At the beginning of 1936 more than 50 per cent of the agricultural indebtedness fell on entailed farms which cannot be mortgaged. In spite of repeated modifications, there has been little progress in the reorganization of agricultural indebtedness set up in the law of June 1, 1933, which provided for a reduction of debts and interest charges by conversions. The aims of the new structure of credit organization are to liquidate old indebtedness, and limit new indebtedness to purely productive outlay. Long-term credits, between six and twenty years' maturity, are restricted to constructional work, land reclamation, or land purchase for younger sons on entailed farms.

In summary, the social and economic aspects of the Nazi agricultural policies dovetail remarkably well, if considerations arising from classical economic precepts about the advantages of world trade are relegated to the background. The various measures which checked the drop of agricultural profits tended not only to reconcile the interests of large landowners with the interests of industrialists, whose profits were likewise protected, but also to develop the nation's independence in foodstuffs. The large estates lent themselves more readily to intensive, scientific, mechanized farming than small tracts, and the agrarian measures taken by the government seem on the whole to have benefited large-scale producers, who were able to take full advantage of the new policy at once, more than the

[25] N. Steinberger, *Die Agrarpolitik des Nationalsozialismus* (Internationalen Agrarinstitut: Moscow, 1935), p. 117.

ordinary run of small farmers. Certain aspects of the resettlement program found support not only from landowners but also from industrialists who saw the possibility of moderating labor unrest by giving it a property interest. The political aim to make Germany independent of foreign food markets has helped to fulfill the economic aspirations of the farmer and at the same time to preserve a bankrupt landed aristocracy. Thus, in the short run, a war economy which serves as a fillip to heavy industry also works to the advantage of certain farmers. From the point of view of the whole nation, however, the effect of these policies is to reduce the standard of living. Military, political, agrarian, financial, and industrial interests of the nation reinforce each other in the development of an autarchic economic regime.

XI

NATIONAL INCOME, CONSUMPTION, AND SOCIAL WELFARE

THE preceding pages have been devoted chiefly to a description of National Socialist credit, public works, and armament policies, and to the mechanism established under Hitler to manipulate the German economy. If the true significance of the relationship between state and economy is to be determined, however, these various regulations must be viewed as a whole. In this connection, the question again arises: Which groups gain economically and which lose?

The National Socialist economic state can be pictured as a national "trust," which combines the economic forces of industry and agriculture, large and small enterprises, capital and labor, and which is enforced by Hitler or his representatives and subordinates, the Ministers of Economic Affairs, Agriculture, and Labor. The problem, however, of which interest groups are the most favored still remains unsolved.

There is one group which appears unequivocally to draw special favors from the Hitler government: the National Socialist party and its affiliated organizations. Many of the privileges connected with the National Socialist party are of an economic character and cannot be set aside as unimportant — preferences in labor placement, distribution of settlement lots, allotment of public works orders, raw materials, etc. This distinction based on party affiliation, however, is essentially political and affects industrialists or bankers as well as farmers, workers, and handicraftsmen, and therefore, does not answer the question: Which group benefits from the Nazi regime?

The following discussion attempts to answer that question as well as to discover whether economic conditions have deteriorated since Hitler came to power, or whether, as the Nazis claim, there has been a steady improvement of living standards. The status of economic conditions is here measured by total national income and per capita income; income received by the worker, the entrepreneur, farmer and rentier; distribution of income according to income classes; per capita consumption of foodstuffs, radios, automobiles, movies, and pleasure trips; and statistics for building construction, education, and sickness. Allowance is also made for tax payments. Considerable attention is devoted to a criticism of the available statistics, explaining where they are incomplete and misleading.

The chief inaccuracies involved in comparisons of total national income for various years result from the considerable changes in prices and in the quality and kinds of goods consumed. The biggest gap in the statistical material is, of course, the amount spent on armaments. According to Hitler, Germany spent roughly 90 billion RM on war preparations before the outbreak of hostilities.

The periods of comparison are a year of prosperity prior to the Nazis, 1929; a year of deep depression, 1932; and the six years of the Nazi regime from 1933 to 1938. Because of the war, official publications and even newspapers are no longer sent regularly to the United States. As a result, the statistics for 1939 are incomplete, while those for 1940 are extremely scarce.

THE SIZE OF THE NATIONAL INCOME

The German statistics show an increase of the national income from around 45 billion RM in 1932 to about 79 billion RM in 1938. In 1938 income was thus above the 76 billion RM of 1929.

Before attempting to make any adjustments for price changes, increased population, and finally for the tremendous

INCOME, CONSUMPTION, AND SOCIAL WELFARE

amounts spent for armed forces and their equipment, we may well ask whether the basic national income figure is a good one. In Germany, as in most other countries, there are no statistics collected directly for the purpose of measuring the size of the national income, and consequently the German Statistical Office must make an estimate from miscellaneous and partial data compiled for other purposes. What is really important in this connection is whether the goodness of the estimate remains fairly stable for all the years used in the comparison or whether it is likely to fluctuate to such an extent that the year-to-year income comparisons are unreliable.

The German Statistical Office defines the national income as the sum of goods and services which are available for consumption and capital formation, and for which a price is commonly paid, after deduction of charges for maintenance, depreciation, and capital losses. And income so defined is analyzed in terms of income paid out and hence represents national income before deductions for taxation and other compulsory payments.[1]

---

[1] The components of national income as computed by the German Statistical Office are as follows:

Included	*Excluded*
a. *Services*: land, labor, and capital. Includes, among other things, incomes paid to all public employees, salaried workers, professional people, and domestic servants. Also services of land and capital owned by individuals, business enterprises, and government corporations.	a. *Services* of housewives to their families and other "honorary" services and the use of durable consumers' goods — for example, the rental value of furniture.
Rental value of homes occupied by owners.	Services representing "mere consumption of property (liquidation of savings for purposes of consumption) and mere transfers of property (loans granted for purpose of consumption, repaid debts, proceeds from the sale of property, legacies, donations, profits from speculation and lottery gains.)"
Certain services of government represented by license fees, etc., paid to government for services directly promoting consumption or production.	
Interest on loans by government to individuals and private businesses.	Income from charities, welfare, relief — both public and private.

The Statistical Office uses for the most part the wage, income, and corporation tax statistics as the basis for determining total national income thus defined.[2] These statistics are unusually comprehensive, but four sources of error tend to inflate the value of income for the boom years of the Nazi regime. They are as follows: failure to include all services;

b. *New capital formation* within Germany. This formation is already reflected in income-service paid out, and the only particular provision in the income estimate is an allowance for undivided profits.

b. New capital formation financed with foreign loans.

c. Income earned abroad as interest on German capital abroad and for German services rendered abroad, e.g., freight charges.

c. Income going abroad, that is, interest on foreign capital invested in Germany, return for services rendered to Germany by foreigners—freighting, etc.

[2] The wage tax statistics are particularly good, since all wage income of less than 8,000 RM must be reported at the source. The extent of incidental income earned by wage earners is estimated for all years on the basis of family-budget investigations.

The goodness of the income tax statistics is indicated by the extent of income liable to tax assessment. Income derived from real estate and enterprises situated in Germany as well as all government salaries, remunerations, and pensions is liable to income taxation. However, only gross incomes above 1,300 RM are assessed (in 1935, incomes above 1,100 RM). Incomes below this amount must be estimated by the Statistical Office. This coverage is better than would appear at first sight because of the manner in which gross income is defined — income *before* deductions for tax exemptions, professional outlays, and interest due on debt. Thus, because of these deductions, an individual may not pay taxes and yet his income is reported.

The income of private corporations, so far as it is distributed in the form of dividends, etc., appears already as personal income. The undistributed profits are estimated with the aid of the corporation tax statistics and the statistics of balance sheet statements. All common-stock companies, joint-stock companies, and limited liability companies are liable to the corporation tax, but mutual insurance companies and non-profit corporations are tax exempt. Public corporations were also exempt until the law of January 1, 1935, which made such corporations subject to the same taxation as business corporations.

Despite revisions in tax laws, coverage of the tax statistics varies only slightly.

INCOME, CONSUMPTION, AND SOCIAL WELFARE

inclusion of the yield from foreign capital invested in Germany which is not permitted to go abroad; under-reporting; and tax evasion. Why each of these involves an "upward" bias for the Nazi period will be discussed next.

Failure to include the services of housewives to their families means that national income will vary with the proportion of housewives' services performed on the market whether or not there has been any actual change in real income. If women's services are performed in industry they are counted as income, but if they are carried on in their homes they are not. The problem can be seen most clearly if we suppose that in one country one million wives remain at home and one million work in industry, and there are no domestic servants; the total income will then be less than if one million wives work in industry and one million women are domestic servants in the homes of the absent wives, despite the fact that the total work being done may be the same in both cases. The emphasis of the Nazis on "kitchen, church, and children" decreased in the first years the total number of women employed outside their own homes, whereas the labor shortage during the later stages of the armament boom brought about an increase in the percentage of women so employed, even as compared to the previous prosperity. The amount of error involved is not great, but it is on the side of underestimating income in the years 1933–35 and exaggerating its growth in the later years, both as compared to depression and as compared with 1928.

Whether failure to deduct for yields of foreign capital invested in Germany, because under the foreign exchange controls it is not permitted to go abroad, involves an error is open to question. These yields were, of course, not included in the 1928–29 income. In so far as the problem is one of determining the income available to the Nazis, these yields should be included, but from the point of view of evaluating the results of Nazi policies it would be desirable to have them excluded.

Tax evasion and under-reporting are probably larger sources of error than those already mentioned. In recent years income is more completely reported because the efficiency and ruthlessness of the Nazis makes evasion extremely difficult. Under-reporting by farmers, moreover, creates a special case. It has undoubtedly been less since the advent of Hitler. A part of farm income — farm products consumed at the farm and theoretically liable to income taxation — is subject only to the farmer's evaluation and not to assessment by the government. In 1928–29 farm products retained for consumption amounted to 27 per cent of the value of agricultural production, whereas in 1938–39 it had declined to 23 per cent.[3] This decline suggests either that tax collections were more rigorous or that there was an increasing market output. In either case part of the increased income must be attributed to more accurate methods of measuring income rather than to increase in farm output.

Having in mind, then, that the national income estimates exaggerate the expansion in income during the period of the Nazis, the next task is to make allowances for price changes. The German Statistical Office does this with the greatest ease by dividing total national income in each year by the cost of living index. The result is called "national income in purchasing power of 1928." The effect of this arithmetical operation is to increase the income of 1938 considerably. National income measured in this way is 68 per cent larger in 1938 than in 1932, and even 26 per cent greater than in the previous prosperity.

The cost of living index, however, is hardly the appropriate divisor for eliminating price changes from the estimate of the total national income. National income is composed of producers' goods as well as consumers' goods; and fluctuations in the prices of all goods and services should be taken into ac-

[3] I.f.K., *Weekly Report*, May 22, 1940, p. 49.

count. In addition, the German cost of living index does not measure accurately changes in the cost of living. Even German economists and statisticians suggest that the official index underestimates the actual extent to which the cost of living has risen since 1932. It is based on a family budget for a working-class family of five persons. Requirements for a German family included in the official index are far too scanty, and there is good reason to believe that the consumer is often driven to buying better qualities of food at higher prices than are allowed for in the index because the cheaper qualities cannot be obtained.[4] In a good many cases there has been a deterioration of quality so that what is nominally the same article may actually be an inferior one. The figures for housing are becoming progressively less representative from year to year, since they are based upon rents payable for dwellings constructed before 1918, whereas in actual fact living accommodation is more and more taking the form of houses built since that year.

The most accurate method of eliminating price changes would be to divide each component of national income by an index of prices suitable for that component — for example, new capital formation in the iron industry divided by an index of iron prices, etc. This obviously is not possible. The next best approach would be to use a general price index as divisor, but unfortunately none is available. There is a wholesale price

[4] The family is supposed to occupy two rooms and a kitchen, require 2 kilograms of margarine per month, 100 kilograms of coal, 150 kilograms of briquettes, 15 cubic meters of gas, and 5 kilowatt hours of electricity. The scale of clothing for a year includes 1 shirt, 1½ undervests, 1 pair of underpants for the man, 3 pairs of artificial silk stockings and 2 pairs of cotton stockings for the women. (Department of Overseas Trade, *op. cit.*, p. 231.)

The weighting of the various groups in the index was as follows: food, 55.4 per cent; rent, 13.1 per cent; heat and light, 4.7 per cent; clothing, 12.9 per cent; and miscellaneous, 13.9 per cent (Reichs-Kredit-Gesellschaft, *Germany's Economic Situation at the Turn of 1937/39*, p. 51).

index, but it is computed in such a way as to be practically useless — it has a pre-war base year of 1913 and an antiquated system of weighting.[5]

The overestimation of real income involved in dividing money income by the German cost of living index or the wholesale price index is clearly revealed if the deflated income figures are compared with indexes for physical volume of production. The increase in the industrial production index since 1928 or 1932 should be greater than that shown by the figures for real national income. In the first place, the production index represents changes in gross industrial output, whereas the national income figure is net income after deduction for necessary replacement and upkeep. In addition, industrial production is only one part of the national output and the portion which has in the main shown the largest increase. Along with the increase in the industrial production index must be averaged the smaller increase in other areas — for example, the smaller increase in agriculture where the volume of production was in 1938 about 15 per cent above the 1928 level and approximately 9 per cent above 1932. In addition, the Nazis have revised the computation of the industrial production index so that undue weight has

[5] The officially computed wholesale price index is composed of 38 price relatives whose base is a weighted arithmetic average of prices during 1913. In order to obtain a representative average it is desirable to select a relatively recent base year, and the German index with its base in a pre-war year falls far short of this requirement. There is, however, an additional objection which is connected with the method of weighting. The wholesale index is made up of seven group indices, each weighted according to the relative consumption of various groups of commodities in the years 1908–12, and there has been considerable change in the relative importance of various groups during the present decade. The *Statistical Year-book* of the League of Nations shows this wholesale price index shifted to 1929 as a base year. What this means is that the index numbers arrived at with 1913 as a base year are divided by the index for 1929. But the difficulty still remains that the 38 commodities selected in a pre-war year as indicators of the change in all wholesale prices may not be representative for the Nazi period.

INCOME, CONSUMPTION, AND SOCIAL WELFARE 203

been given to heavy industry.[6] Notwithstanding these factors, the increase in the index of industrial production was identical with the increase in national income divided by the cost of living index: in 1938 the index was 68 per cent above the 1932 figure and 26 per cent above the previous prosperity in 1928. It was even less than the increase in national income deflated by the wholesale price index, for "real" income in this case was 34 per cent greater in the Nazi boom than at the pre-Nazi peak.

In view of the limitations in using already calculated indexes for the purpose of eliminating price changes, the author has constructed a new index from the available individual series of various wholesale prices with 1925–34 as the base period.[7]

[6] The production index includes 69 series, representing about 66 per cent of the net industrial output. It is calculated by means of a weighted arithmetic average; the weighting is based on the total number of workers employed and on the total horsepower installed in each industry. The weights assigned to the individual industries have been revised on the basis of the provisional results of the 1933 census of industrial establishments. The weighting of the four main groups, however, is based on the net total output in 1927–29 of all industries represented by each of these groups. The index was revised in 1927, 1928, 1931, 1933, and 1935. Revisions have involved for the most part an increase in the number of series covered. In 1927, 12 per cent of industrial production was included; in 1929, 25 per cent; in 1931, 30 per cent and in 1933, 60 per cent.

After 1935, chemicals, gas, electricity, and water-works series are included, and extra weighting was given to the iron, steel, metal, and electrical machinery series. The effect of these changes is to make the index unusually sensitive to the war activity. Also included for the first time are food-products series. This, too, makes the index sensitive to war activity, since food-products are largely canned goods useful as reserves. Moreover, within the textile series greater weight has been given to individual figures which reflect increased production of army uniforms. Since March 1935 the Saar has been included in the index, but not in the base. (I.f.K., *Vjh.*, 6/1/A (1931); 7/4/A (1933) and I.f.K., *Wochenbericht*, 1935, no. 24, p. 19.)

[7] This new index was constructed along lines suggested by Colin Clark in *Conditions of Economic Progress* (London, 1940), p. 94. The weights used were the estimated values of consumption shown in the *Wirtschaftsrechnung* of 1927–28 and values of investment from estimates of the same

THE STRUCTURE OF THE NAZI ECONOMY

Using this index as a price deflator, the increase in national income is less than it was when other indexes were employed: the 1938 national income at 1925–34 prices is 20 per cent above the pre-Hitler prosperity and 61.5 per cent above the 1932 depression. Per capita income at 1925–34 prices shows an increase of 13 per cent since 1929 and approximately 23 per cent since 1932 (see Table 19).

The proportion of this expansion in "real" income which was available for civilian consumption and investment may now be estimated. Hitler has said that Germany prior to the outbreak of war spent 90 billion RM on armaments. This expenditure, amounting to 20 per cent of total income during six and one-half years, is equivalent to an arrangement under which one working day out of every five is devoted, without compensation, to munitions manufacture. If the yearly income had stayed at its 1932 level of 45 billion RM, 300 billion would have been produced in six and one-half years. As it was, national income for that period totaled 433 billion, an increase

---

date by the Institut fuer Konjunkturforschung. The weights were as follows:

Consumption Goods	Per cent
Food	36
Rent	10
Clothing	13
Fuel	4
Misc.	26
Investment Goods	
Building	5
Agricultural equipment	1
Industrial equipment	5
	100

The price indexes thus calculated with 1925–34=100 were as follows:

1913	70.7	1929	109	1934	86
1925	104	1930	108	1935	88
1926	104	1931	99.5	1936	90
1927	106	1932	87.5	1937	93.4
1928	109	1933	86	1938	95

for the whole period of 133 billion RM. Hitler's armament estimate of 90 billion RM represents two-thirds of this increase. Thus, at the most, per capita income, including both civilian consumption and investment, has increased only eight per cent since the depths of the depression.

The Nazis have thus had remarkable success in achieving military goals, but the results of their policies from the point of view of civilian needs are less happy. Increasing armament activity, along with autarchic policies which tend to lower productivity, have, however, not depressed the standard of living below depression levels. This fact may be accounted for in good part by increased employment and longer hours. Greater utilization of productive capacity as well as increase in the size of plants also contributed to balance offsetting tendencies with regard to efficiency and productivity. While the considerable growth in size of corporations and the numerous mergers and consolidations in many cases meant merely increased centralization of financial control without any improvement in the organization of production, the stable price policy tended to force firms toward vertical or horizontal integration. Increased efficiency, together with more nearly optimum utilization of plant, was the main factor contributing toward larger profits in a situation where selling price was fixed and the cost of raw materials — imported or *ersatz* — was rising. The Estate of Industry and Trade and the Reich Food Estate likewise encouraged rationalization.

The exact role played by increased mechanization cannot be determined until the 1939 Census of Manufacturers is available — the last census was taken in 1933 — but the considerable increase in the use of electric power suggests that this factor was very important. The very nature of a war economy leads to increased mechanization, particularly if increased armaments are produced at the expense of consumers' goods, usually less mechanized. The decline in road construction after the outbreak of hostilities had the same effect. The shift into more highly mechanized industries has undoubtedly bolstered up the decline in the tempo of increased productivity which

TABLE 19
NATIONAL INCOME

	Unit	1929	1932	1938
National income in current prices	Billion RM	75.9	45.2	79.7
National income in 1928 prices	Billion RM	74.8	56.8	96.0
National income in 1925–34 prices	Billion RM	70.0	52.0	84.0
(At least ⅔ of increase in national income goes for armaments)				
Per capita income, 1925–34 prices	RM	1089	998	1226
Index of physical volume of industrial production *		100(1928)	58	126
Index of physical volume of agricultural production *		100(1928)	106	115
FACTORS CONTRIBUTING TOWARD INCREASED INCOME				
Use of electric power †	Index (1928=100)	117.9	90.4	229
Utilization plant capacity ‡	Per cent	67.4	35.7	81.0
Employment §	Million	17.9	12.6	19.5
Industrial workers ‖	Million	6.2	3.7	7.3
Total man hours worked in *industry*	Billion hours	14.6	7.9	17.2
Total man hours worked	Index (1928=100)	104	55	118
Average size of corporation, measured by capital stock ¶	Million RM	2.0	2.3	3.4

Data is taken from *Stat. Jahrb.* and *W. und S.* unless otherwise indicated. See text for description of price index, 1925–34 = 100.
  * I.f.K., *Statistik des In- und Ausland*, 1939/40, Heft 2.
  † I.f.K., *Weekly Report*, August 26, 1939.
  ‡ Statistics for utilization of plant capacity are given only for 1929–35, I.f.K., *Konjunkturstatistiches Handbuch*, 1936, p. 25. The *Handbuch* defines utilization of plant capacity in the following manner: total number of man hours actually worked divided by total number of man hours possible when plant is working full time at peak load. The increase for 1938 was computed by multiplying the 1935 percentage by the increase in the index for total number of hours worked, increase of 1938 over 1935. This increase was then added to the percentage figure given for 1935 in the *Handbuch*.
  § I.f.K., *Vjh.*, 1939/40, Heft 1, p. 5.
  ‖ I.f.K., *Wochenbericht*, 1939, No. 9, p. 46.
  ¶ *Vjh. Stat. d. D. R.*, 1939, I, p. 121.

occurred as labor and material shortages appeared. Increased productivity was also to be expected from the limitations placed on labor turnover as a result of universal labor conscription in 1938. Mass production was encouraged by such measures as restriction of the variety of types in the motor vehicle industry. The changes which occurred in broad industrial divisions were also conducive to greater efficiency. A shift from agriculture to industry, as represented by changing relative shares in the total number of gainfully occupied, has in all countries been accompanied by increased productivity. Since 1895, industry in Germany has bulked larger than agriculture and, despite

TABLE 20

GAINFULLY OCCUPIED IN OLD REICH, 1933–39

(*Per cent of total gainfully occupied*)

	1933	1939
Agriculture	24.3	18
Industry	34.8	41
Trade and commerce	15.5	15.8
Government service	7.1	10.0
Domestic servants	3.3	2.2
Independent merchants	15.0	13.0
	100.0	100.0

Computed from data in *Stat. Jahrb.*, 1938, p. 25 and Deutsche Bank, *Wirtschaftliche Mitteilungen*, September 30, 1940, pp. 136–137.

Nazi attempts to increase the number of farm laborers, the movement away from agriculture has merely slowed down (see Table 20). The compulsory liquidation of small shops released merchants for industrial work of higher productivity. The decline in domestic servants had the same effect. These two factors, however, were offset by greater bureaucratization.

THE STRUCTURE OF THE NATIONAL INCOME

Statistics of the division of national income among workers, farmers, business men, and other groups are shown in Table 21 and reveal that the relative position of these groups has changed since the Nazis came to power. The most striking change is the increased share going to property and the decreased share represented by earned income. From these figures

TABLE 21

STRUCTURE OF THE GERMAN NATIONAL INCOME, 1929–38

	1929	1930	1931	1932	1933	1934	1935	1936	1937	1938
Total national income (billion RM)	75.9	70.2	57.5	45.2	46.6	52.7	58.6	65.0	71.0	79.7
				PER CENT TOTAL NATIONAL INCOME						
Agriculture and forestry	7.2	7.0	7.6	8.2	8.3	9.4	9.6	8.5	7.9	7.3
Wages	31.4	29.9	28.7	26.1	26.4	28.2	28.5	28.5	29.5	30.2
Salaries of insured employees	10.3	11.2	10.1	13.0	12.4	12.1	12.3	12.6	13.0	12.5
Other salaries	15.0	15.7	19.3	17.8	17.0	15.1	14.3	13.2	12.1	11.5
Social insurance, pensions and annuities	12.1	14.2	17.5	20.7	18.3	14.9	13.1	11.4	9.9	9.5
Total from salaries, etc.	68.8	71.0	75.6	77.4	74.1	70.3	68.2	65.7	64.5	63.7
Profits	15.5	14.2	13.0	13.3	13.8	13.7	14.5	16.0	16.9	18.6
Undistributed profits	1.2	0.6	−1.7	−1.0	0.4	1.3	2.0	2.8	3.1	4.3
Interest and dividends	4.3	4.7	5.6	5.1	5.2	4.9	4.4	4.1	4.0	3.7
Rent	1.1	1.3	1.6	1.7	1.5	1.5	1.4	1.5	1.4	1.4
Total from property	22.1	20.8	18.5	19.1	20.9	21.4	22.3	24.4	25.4	28.0
Employer's contribution to social insurance not elsewhere reported	3.2	3.4	3.8	3.8	3.6	3.8	3.7	3.7	3.7	3.6
Indirect taxes not reported in entrepreneurial income	4.8	5.7	6.4	5.7	5.3	4.4	3.9	3.5	3.2	2.9
Government income from enterprises, land and capital	3.2	3.3	2.1	2.2	2.0	1.9	2.0	2.4	2.2	1.8
Deduction for double counting	−9.3	−11.2	−14.0	−16.6	−14.2	−11.2	−9.7	−8.2	−6.9	−7.3

SOURCE: Computed from data in *Stat. Jahrb*, 1938, p. 560, and *W. und S.*, XIX (1939), pp. 298, 705.

INCOME, CONSUMPTION, AND SOCIAL WELFARE

it appears that entrepreneurs and rentiers fared better than workers. In 1929 total income from various kinds of property made up 22 per cent of the national income, while in 1938 it represented 28 per cent. In contrast to this, earned income dropped from its level of approximately 69 per cent of total income to approximately 63 per cent in 1938.

The decline in the share going to workers is largely a result of extremely low wage rates. Wage rates were maintained until 1930, long after the boom had ended, but were drastically reduced in 1931 and 1932. After the Nazis, wage rates were about 22 per cent less on the average than at their previous peak. Wage rates fell slightly during 1933 and were stabilized at their lowest level. Although total hours worked in 1938 exceeded the number of man hours during 1929 by about 15 per cent, the share going to workers still remained below the level of 1929. The worker's position is not significantly changed even when allowances are made for relative tax burden (see Table 22). As far as taxes and compulsory and "voluntary" contributions are concerned, an average worker spends about 22 per cent of his income in this way.[8] In 1937, for the community as a whole, taxes and compulsory contributions made up about one-third of national income.[9]

It is, of course, true that not all of this expenditure is a net loss to the worker. The Labor Front through its "Strength through Joy" department has performed the services shown in Table 23.

Although the Labor Front has provided about one-third of the workers in each year with extraordinarily cheap vacation trips, and facilitates admission to theaters and concerts and sports, its activities are insignificant compared to the great bulk of expenditure on armaments. Labor Front services,

[8] *Einzel.*, 1937, Nr. 35, p. 177.
[9] Rudolf Brinkmann, *Staat und Wirtschaft* (Stuttgart and Berlin, 1938), p. 23.

210   THE STRUCTURE OF THE NAZI ECONOMY

moreover, serve the double purpose of diverting the worker's attention from political matters and the shortcomings of the Nazi system and at the same time of directing his expenditures where they will not require raw materials needed for armament.

The remaining share of national income, not yet discussed,

TABLE 22

Taxes and Compulsory Expenditures of a Working-Class Family in 1936

(*Unit: Per cent of individual income*)

	Family with 2 children Income 2,400 RM (Per Cent)	Income 4,200 RM (Per Cent)
Income or wage tax	2.08	3.94
Poll tax	1.16	.83
Church tax	.20	.39
Excise tax	3.44	1.45
Tariff duty	.62	.47
Turnover tax	2.60	2.52
Social insurance		
Sickness	3.90	2.82
Unemployment	3.45	2.29
Accident	2.60	5.17
Voluntary Contributions		
Membership in Nazi party, Labor Front, National Socialist Welfare Organization, Reich Air Defense League and the Association for Germans Abroad and other affiliated groups	3.41	3.64
Total	22.00	18.35

Source: *Einzel.*, 1937, Nr. 35, p. 177.

is agriculture. Its relative position is slightly better than in the previous pre-Nazi prosperity and considerably improved over the depression level. The Nazis gave particular favors to this section of the economy during the first years of their regime. Agricultural prices were protected from foreign competition by protective tariffs and the government deliberately aimed its price policy to close the gap between industrial and agricultural prices. As national income has expanded, however, the share going to agriculture has not kept pace. Diminishing returns explain this trend in part, but the most significant factor was the shift of the economy into large armament production.

TABLE 23
"Strength through Joy" Activities, 1934–37

Vacation trips on land and sea, participants	19,490,000
Sports activities, participants	18,000,000
Week-end trips, participants	2,960,000

Source: *Stat. Jahrb.*, 1938, p. 638.

DISTRIBUTION OF INCOME

General interest in the distribution of income according to size in Nazi Germany arises not only from its close relation to spending-saving patterns which influence the course of the business cycle but also from its relevance to the question of who benefits from the regime.

The German data on the distribution of income according to income classes are fairly comprehensive. The total number of those included is equal to the total number of the gainfully employed minus family members in the family business, plus persons living on the earnings from their investments; and it amounts to more than thirty million individuals in any one year.[10]

The frequency distribution of the various income classes as tabulated by the German Statistical Office reveal extraordinary skewness, such as is customarily found in income data. This skewness not only robs the average and other simple statistical summarizations of their significance but also prevents the charting of these data by the simple methods ordinarily used for frequency series.

[10] For a full discussion of the statistical data, see Maxine Sweezy, "Distribution of Wealth and Income under the Nazis," *Review of Economic Statistics*, November 1939.

212 THE STRUCTURE OF THE NAZI ECONOMY

For such skewed series the graphic method devised by Pareto is often helpful, and it is used here as the general means of representation and analysis. The application of the method does not necessitate any implication as to a "law" of distribution for values of income outside the range covered by the observed data; it is used solely to describe the distribution within the range of the given statistics. According to this method, the frequencies are cumulated, from the highest class, 100,000 RM,

TABLE 24
DISTRIBUTION OF INCOME IN GERMANY ACCORDING TO SIZE

Lower Limit Income Class (1,000 RM)	\multicolumn{5}{c}{CUMULATED FREQUENCIES (1,000)}
	1926	1928	1932	1934	1936
1.2	10,847	13,180	9,863	11,664	11,492
3	2,348	3,239	2,279	2,563	2,504
5	863	1,248	716	821	1,002
8	333	469	196	256	367
12	161	218	103	125	189
16	97	132	60	74	119
25	44	59	26	32	57
50	13	17	7	9	18
100	4	5	2	2	5
Average Inequality	.549	.552	.500	.491	.599

Computed from data in *Stat. Jahrb.*, 1937, p. 535; and *W. und S.*, 1939 (XIX), pp. 922 and 961.

toward the lowest, 1,200 RM. A chart is constructed with horizontal measurements representing size of income and vertical measurements representing cumulated frequency. Both scales are logarithmic. The cumulated frequencies are contained in Table 24; the corresponding Pareto-type curve is shown in Chart III.

The plotted curves are, moreover, so nearly straight that the average inequality of the whole range of incomes can be inferred from the direction of a straight line joining two widely separated points of the plotted curve.[11] The points chosen are for statu-

[11] If the frequencies for this lowest class interval, under 1,200 RM, are plotted, the curve reveals upward convexity, and it therefore seemed

CHART III
Distribution of Income in Germany

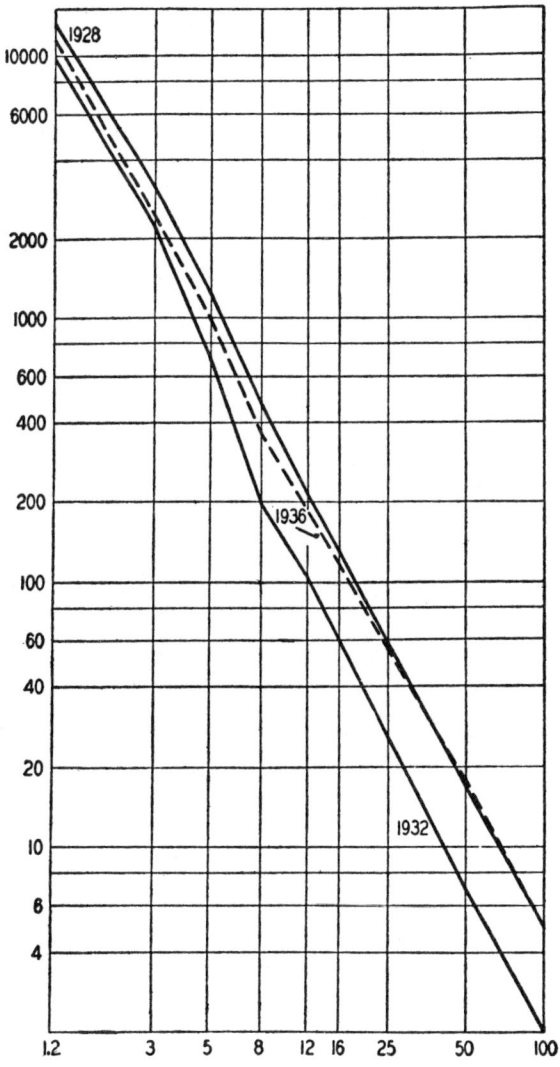

Horizontal scale represents size of income; unit, 1,000 RM. Vertical scale represents cumulated frequency; unit, 1,000. Both scales logarithmic.

tory net income of 1,200 RM and 100,000 RM. The slope of this line gives the ratio of the percentage change in number of individuals falling in various income classes to the percentage change in size of income class. An increase in steepness of the slope of the straight line means a decrease in inequality, for, if the percentage change in the number of individuals is relatively large for a given percentage increase in individual income, people are closely alike with respect to size of income. This reverse relationship between the slope of the line and the degree of inequality, and correspondingly between the direction of change in slope and the direction of change in inequality, is somewhat awkward and has frequently led to a misunderstanding of the direction of the change taking place in income distribution. To avoid this confusion, an average index of inequality has been computed in which the slope is inverted. By inverting the usual measurement of the slope, an increase in the index represents an increase in inequality.

There was a slight tendency for the slope to be greater in depression than in prosperity, as can be seen from a comparison of the curve for 1932 with those for 1928 and 1936. Average inequality was not only greater during prosperity than in depression, but it was also larger in the Nazi boom year of 1936 than in 1928. Comparisons of this sort must be qualified by due consideration of the somewhat treacherous way in which the Pareto method of charting condenses the statistical evidence. Hence, small changes in the index — for example, the slight decrease in average inequality in 1934 — cannot be considered significant.

A more detailed picture of the changes in inequality can be obtained from a close examination of the steepness of slope

---

wise to discuss distribution of income below 1,200 RM in connection with wages and salaries for which there are frequencies related to more finally graduated class intervals. Probably most of the income in this lower bracket represents wages and salaries, although this is not completely true.

between different points of the curves for 1928 and 1936. The slope of the line between the 50,000 RM and the 100,000 RM points is slightly steeper for 1936 than for 1928; on the other hand, it is less steep in 1936 between 8,000 RM and 50,000 RM. This means that the percentage of individuals receiving very large incomes increased in 1936 in comparison with 1928, while the percentage number of medium-sized incomes diminished. There is little variation from year to year in the percentage number of lower income receivers.

Unfortunately, the data available on the lower income classes when plotted show such decided curvature that it is not possible to compute an index of average inequality by inverting the slope of a straight line.[12] For this reason, the very small income groups, as shown by the statistics for wages and salaries, are presented in the form of percentage number of individuals falling in various wage or salary classes. The income data as reported by Invalid Insurance and the Office Employees Insurance Office include not only industrial wage earners, but also workers receiving remuneration in agriculture and trade — young clerks, journeymen, apprentices, and domestic servants. Included in wage income in addition to money wages are industrial bonuses received by wage earners, payments in kind, and all other items, with the exception of pure gifts, which the wage earner receives in place of wages or in addition to them. Salaries cover all clerical employees, including managers who do not earn more than 7,200 RM annually and are under sixty years of age.

Increase in the percentage of workers in the extremely low wage groups is striking. Whereas the number receiving a weekly wage of only twelve RM or less amounted to 15.8 per cent in 1929, it came to 22 per cent in 1938. There was little change in the upper wage brackets except for the highest class

[12] These charts are contained in Sweezy, *loc. cit.*, which has a full discussion concerning the adequacy of the wage and salary statistics.

interval, where the percentage for 1938 was below that for 1929 (see Table 25).

Distribution of salaries showed a different pattern. There

TABLE 25
DISTRIBUTION OF WAGE INCOME
(*Unit: Percentage of total number of wage receivers*)

Size Class of Weekly Wages	1929	1932	1938
To RM 6	3.5	3.9	4.0
Over 6 to 12	12.3	18.3	18.0
Over 12 to 18	16.5	22.7	12.1
Over 18 to 24	13.0	16.3	11.3
Over 24 to 30	8.8	10.7	10.3
Over 30 to 36	8.1	9.0	10.9
Over 36	37.8	19.1	33.4

SOURCE: I.f.K., *Vjh.*, 1939/40, Heft 1, p. 12.

was a noticeable increase in the number of salaries above 500 RM, whereas the percentage for the lowest salary receivers was somewhat less (see Table 26).

TABLE 26
DISTRIBUTION OF SALARIES
(*Unit: Percentage of total number of salary receivers*)

Monthly Salary	1929	1932	1938
To RM 50	13.7	13.6	11.9
Over 50 to 100	14.5	21.1	14.6
Over 100 to 200	30.2	31.9	29.5
Over 200 to 300	18.8	17.0	20.4
Over 300 to 400	11.4	8.6	10.6
Over 400 to 500	6.3	4.2	5.6
Over 500 to 600	2.5	1.8	4.1
Over 600	2.6	1.8	3.3

SOURCE: I.f.K., *Vjh.*, 1939/40, Heft 1, p. 12.

DISTRIBUTION OF WEALTH

The Pareto method has been applied to the distribution of wealth according to size classes, and the results are shown in Chart IV and Table 27. The data are derived from the prop-

CHART IV
Distribution of Wealth in Germany

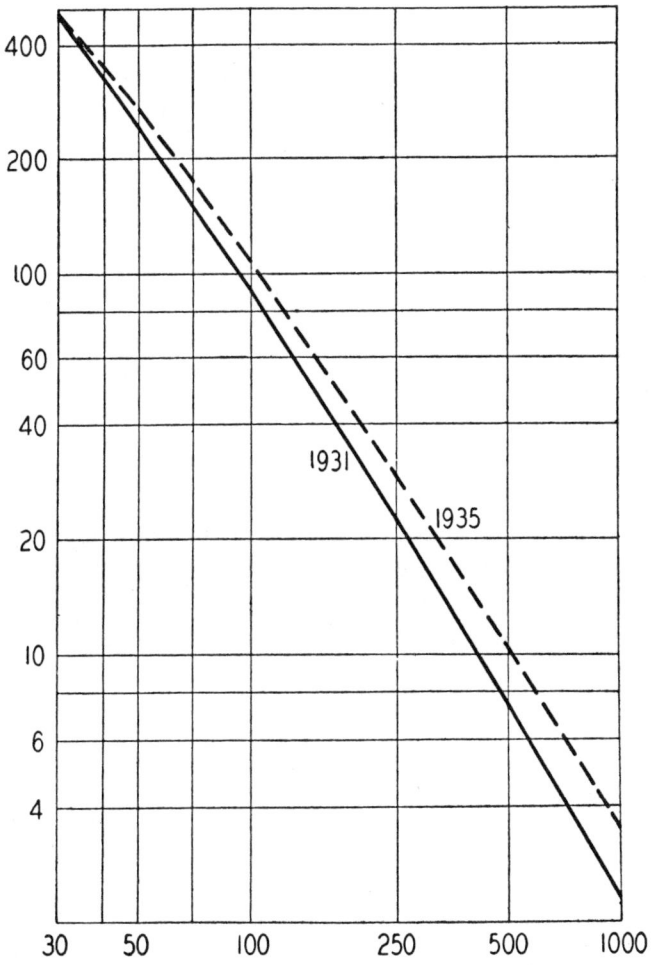

Horizontal scale, size classes; unit, 1,000 RM. Vertical scale, cumulated frequency; unit, 1,000. Both scales logarithmic.

erty tax statistics and are remarkably good for the range of total property covered from the 30,000 RM size class to the 1,000,000 RM class. Total property, which is the criterion of size class, is defined in the law in the following manner: Gross property — composed of property in agriculture and forestry,

TABLE 27
DISTRIBUTION OF WEALTH IN GERMANY, 1931 AND 1935

Lower Limit of Class (RM 1,000)	1931 Number	1931 Cumulated Number	1935 Number	1935 Cumulated Number
30	234,548	481,647	131,780	488,052
40	......	......	82,243	356,272
50	155,676	247,099	97,242	274,029
70	......	......	67,868	176,787
100	68,925	91,423	79,510	108,919
250	15,048	22,498	19,056	29,409
500	5,126	7,450	6,790	10,353
1000	2,324	2,324	3,563	3,563

Average inequality: 1931 .... .649
1935 .... .721

Computed from data in *Stat. d. D. R.*, Band 519, pp. 10 and 13. Income does not include tax-exempt amounts.

real estate, business undertakings and other property, comprising mainly stocks and bonds — minus debts, other liabilities, and tax exemptions, constitutes the total property.[13] The total property of all physical persons who normally reside in Germany is liable to the tax.

Both curves for 1935 and 1931 reveal a striking approach to

[13] Stocks are assessed at their full market value (*Stat. d.D.R.*, Band 519, pp. 3–10). The general tax exemptions for 1935 were as follows: 10,000 RM for the taxpayer; 10,000 RM for his wife; 10,000 RM for each minor dependent child; and an additional tax exemption of 10,000 RM if the taxpayer was over sixty years of age, or if he would probably be disabled for not less than three years, provided his last annual income did not exceed 3,000 RM. In 1931 there was a uniform exemption of 20,000 RM. In 1935 the following items were also exempt from the property tax: dwelling houses, small apartment and one-family houses constructed during 1933–35, and property exempt because of double taxation.

linearity, with average inequality greater in 1935 than in 1931. The two years are not perfectly homogeneous because of slight changes in the amount of tax exemption permitted in the law,[14] as well as some modification with respect to prices and hence valuation. These price changes, however, were not great because of the stable price policy pursued by the Nazis. Moreover, even though these rather small price changes were not uniform, and hence affected different forms of property differently, the curve would merely shift, provided the various total properties were made up of similar forms. It is also unfortunate that the data are available only for 1931 and 1935, since the greater equality of 1931 may be the result of the depression. The outstanding feature of the picture drawn here is the fact that the Nazi economy can hardly be looked upon as a "socialist" regime so far as ownership of property is concerned. Whatever may have been the specific reasons, the average inequality in the distribution of wealth was greater in 1935 than in 1931.

The increase in the inequality of wealth contemporaneously with increased inequality of income is, of course, not decisive in determining all the causal factors at work. According to Pareto, the distribution of income is determined not by the economic structure of society and by its institutions but by the distribution of certain natural qualities inherent in men. He based his conclusion on the results of a statistical investigation which revealed a striking stability of the income curve in different epochs and places. But, not long after its enunciation, variations in slope of the Pareto line from time to time, and from place to place, and deviations from linearity were recognized as sufficient to cast doubt upon the accuracy and universality of the "law." [15] A recent study of inequality of incomes

[14] *Stat. d. D. R.*, Band 519, pp. 3–10.
[15] Studies of Sir Josiah Stamp, *Wealth and Taxable Capacity* (London, 1930), p. 87, and Arthur Bowley, *The Changes in the Distribution of the National Income, 1890–1913* (Oxford, 1920), p. 27, show a striking stability of the income curve for England. But these studies pertain to a

in Prussia for selected years between 1821 and 1928 revealed that very distinct shifts occurred in periods of economic or political unrest.[16] The statistics shown here also contradict Pareto's contention that the distribution of income is hardly affected by changes in economic structure or institutions, and indicate that inequality increased under the National Socialist regime. The average inequality during a prosperous year under the Nazis was greater than that for a prosperous year before the dictatorship.

The increase in inequality of income, however, has significance over and above its relevance to the characteristics of the institutional changes under Hitler. Generally speaking, the more unequally distributed income is, the greater will be the amount saved by the community at each level of total national income. The increase in inequality of incomes thus assists the more direct attempts of the Nazis to cut consumption drastically in order that the remainder of the national income may be devoted to armaments.

LIMITED RECOVERY OF CONSUMPTION

The most important commodities involved in the limited recovery of consumption will be discussed next. As Table 28

---

period of flourishing capitalism and therefore tell us nothing about the influence of different social institutions upon distribution. Gustav Schmoller studied the income distribution in Bale, Frankfort, and Augsburg in the fifteenth century and in Saxony and Oldenburg in more recent years ("Die Einkommensverteilung in alter und neuer Zeit," *Jahrbuch fuer Gesetzgebung, Verwaltung und Volkswirtschaft*, 1895, p. 1067). He also concluded that differences in personal qualities were the basic cause of inequality but admitted that chance, violence, the economic situation, and social institutions might have some influence. The statistics for the fifteenth and sixteenth centuries are too meagre to be conclusive, and lack of homogeneity may explain Schmoller's results rather than any fixed system of causation.

[16] Costantino Bresciani-Turroni, "Annual Survey of Statistical Data: Pareto's Law and the Index of Inequality of Income," *Econometrica*, April 1939.

reveals, increased consumption of foodstuffs was dominated by the self-sufficiency program. Foods which could be easily pro-

TABLE 28

FOODSTUFFS AVAILABLE FOR CONSUMPTION

	Kilograms [2.2 lbs.] per Capita		Percentage Increase over 1932
	1932	1938	
Meat	42.1	47.8	13
Eggs	138.0	124.0	−10
Fish	8.5	11.9	40
Cheese	5.2	5.6	8
Milk	105.0	112.0	7
Butter	7.5	8.8	17
Margarine and vegetable fats	11.3	8.7	−26
Lard	8.5	8.4	−1
Total fats	(27.3)	(25.9)	(−5)
Wheat flour	44.6	51.9	16
Rye flour	53.5	53.0	−1
Rice	2.9	2.4	−17
Potatoes	191.0	183.0	4
Total starch	(292.0)	(290.3)	(−1)
Legumes	2.0	2.3	15
Green vegetables	47.3	47.0	−1
Citrous fruits	8.0	7.0	−12
Other fruits	30.8	20.3	−34
Total fruits and vegetables	(88.1)	(74.3)	(−16)
Sugar	20.2	24.3	20
Cocoa	.89	.91	2
Coffee	1.6	2.31	44
Beer	51.4	68.6	33
Wine	4.0	6.1	53
Brandy	.6	1.2	98
Tobacco	1.6	1.9	18
Cigars	85.0	131.0	54
Cigarettes	483.0	676.0	40

SOURCE: *W. und S.*, XIX (1939), p. 463.

duced at home were plentiful, as well as those which could be obtained through barter or clearing agreements. Frequently this led to somewhat surprising results: while green vegetables, fruits, and fats declined below depression consumption, brandy, tobacco, wine, and coffee showed enormous increases. The decline in eggs available for consumption has been balanced by increased consumption of fish, meats, milk, and cheese. Starches continue to be the large item in the German diet that they were in the depression. These figures are in some respects deceiving as far as actual consumption is concerned, for they show amounts available for consumption from domestic production and imports. They thus fail to indicate how much of the available consumption is not consumed but kept for war reserves. Changes in quality of food are also hidden. For example, there has been a marked shift from fatty cheese to the less nutritious lean cheese. Part of the rise in flour available for consumption was created by raising the extraction rate of flour from grain.

These figures are in general substantiated by indexes on the volume of agricultural production. Although Germany's food imports were approximately 27 per cent less in 1938 than in the pre-Nazi prosperity, the volume of agriculture had increased only 15 per cent. At the same time, however, there had been an increase in population. It is thus evident that recovery in consumption must be accounted for largely outside of nutritious foodstuffs.

The need for uniforms of all kinds contributed greatly to the demand for shoes and textiles, the expenditure on which increased from six billion RM in 1932 to ten billion RM in 1938.[17] An expansion of 10 per cent in 1938 as compared even to the previous 1929 prosperity in the retail sale of furniture is in part symptomatic of limitations placed on consumption in other areas.[18] Because domestic production is so largely con-

[17] I.f.K., *Vjh.*, 1939/40, Heft 1, p. 19.
[18] *Ibid.*, expenditure for furniture and household needs were as follows:
1932............2.4 billion RM
1929............4.5 billion RM
1938............5.0 billion RM

INCOME, CONSUMPTION, AND SOCIAL WELFARE 223

centrated on armament, public building, and road construction, and priority is given to imports of raw materials for these purposes, it is not possible to buy goods in the variety, quantity, and quality desired. There was also a sharp rise in the number of passenger automobiles owned and registered in Germany. At the middle of 1939 there were 1,486,451 passenger auto-

TABLE 29
BUILDING CONSTRUCTION IN GERMANY, 1929–38

	Number of New Dwellings	Number of Reconstructed Dwellings	All Other Building Construction 100 cbm.
1929	315,703	23,099	60,392
1930	307,933	22,327	69,832
1931	231,342	20,359	50,385
1932	131,160	27,961	35,563
1933	132,870	69,243	34,343
1934	190,257	129,182	39,169
1935	213,227	50,583	56,835
1936	282,466	49,904	72,413
1937	308,945	31,447	76,509
1938	276,276	29,250	77,060
1939		220,000	
1940		100,000	

SOURCE: *W. und S.*, 1938 (XVIII), 428, and 1939 (XIX), 422; I.f.K., *Wochenbericht*, December 28, 1940, p. 153.

mobiles as compared with only 486,000 six and one-half years earlier.[19]

Other luxury commodities have also experienced a rise in demand. The number of radios owned in Germany rose from 4.31 million at the end of 1932 to 11.5 million in July 1939.[20] Larger incomes since the depression have likewise brought an increase in attendance at motion picture theaters and a 50 per cent rise in the number of trips taken on German railways.[21]

War preparation, however, prevented adequate measures to relieve the housing shortage which had developed from the

[19] *W. und S.*, XIX (1939), 614.
[20] *W. und S.*, XIX (1939), 512.   [21] *W. und S.*, XIX (1939), 438.

depression. Although the number of private houses constructed since the depression has increased, it is still below the level for 1929. Grants in aid by the government began to taper off after 1934, housing schemes yielding priority to munitions. It is estimated that half of the annual marriages in Germany indicates the potential demand for additional housing accommodations without allowing for accumulated shortages from previous years. On this basis the provision for housing has lagged far behind requirements, for the estimated normal increase in 1936 alone was 300,000 families (see Table 29).[22] Private construction of houses has been limited by state control, while government production has not filled the needed gap. The number of new dwellings in 1929 exceed those in 1938 by almost 40,000. On the other hand, all other construction was 25 per cent greater in 1938 than in 1929.

SOCIAL WELFARE

General economic well-being is also reflected in the provisions for social insurance, education, and public health; and these social conditions may be used to confirm the general conclusions drawn from the size and structure of the national income and the extent of consumption. Public health is particularly indicative. It is chiefly determined by three factors: the state of the medical sciences, the general sanitary situation, and the character of general working and living conditions. Progress in the first two factors is usually very slow and is seen only by observing development over long periods. Changes in general working and living conditions, on the other hand, find their expression in the improvement or deterioration of public health during short periods, especially if these changes are very marked.

Statistics on reported sick cases for various years are contained in Table 30. Those types of sickness which can be

[22] Department of Overseas Trade, *op. cit.*, p. 204.

TABLE 30
Cases of Sickness Reported

	1928	1932	1933	1934	1935	1936	1937 *
I. General							
Spotted fever	1	3	4	1	1	....	....
Smallpox	2	3	....	....	....	....	....
Scarlet fever	95,909	55,923	79,830	114,923	112,509	124,506	117,544
Diphtheria	46,905	65,414	77,340	119,103	133,843	148,062	146,733
"Genickstarreubertragbar"	823	494	617	1,100	1,362	1,354	1,574
Infantile paralysis	975	3,869	1,318	1,768	2,153	2,234	2,723
Typhoid	11,881	8,756	6,188	7,105	5,858	6,085	6,806
Poisonous infections	1,361	1,707	1,565	1,513	1,751	3,674	....
Dysentery	3,395	5,058	2,685	3,513	3,430	5,057	7,545
Mad-dog bites	318	445	105	65	78	98	80
Spleen disorder	252	83	84	69	86	73	90
"Trichinose"	1	1	2	44	12	....	....
II. Factory sickness and accidents according to social insurance reports							
Reported cases	4,332	6,671	7,133	7,664	8,980	10,570	....
Cases receiving insurance	417	1,742	1,258	1,043	1,161	1,432	....

* Preliminary figures. None available for factory sickness.
SOURCE: *Stat. Jahrb.*

attributed largely to a run-down physical condition created by poor diet, worry, and overwork show a spectacular rise in the last five years. Cases of dysentery and poisonous infections more than doubled, while the number of persons suffering from diphtheria almost trebled.

The increased number of factory illnesses and accidents is likewise astonishing. The number of reported cases in 1936 had increased to more than two and one-half times the number

TABLE 31
PENSIONS PAID UNDER SOCIAL INSURANCE PROVISIONS
(*In 1,000 RM*)

	1928	1933	1934	1935	1936	1937
Disability	742,275	865,605	893,050	911,618	920,005	925,861
Sickness	7,359	3,552	3,446	3,298	3,100	2,881
Old Age	21,712	10,500	9,143	7,838	6,663	5,492
Widows	97,750	130,781	135,115	141,766	146,560	151,660
Orphans	9,356	48,213	47,428	44,424	40,936	38,303
All other	113,430	27,597	30,066	30,242	31,030	31,841
Total	991,882	1,086,167	1,118,248	1,139,186	1,148,294	1,156,038

SOURCE: *Stat. Jahrb.*

in 1929,[23] while the number of insured accident cases increased to more than three times.[24]

"Speeding-up" frequently means more accidents and may be responsible for the considerable increase during 1936. Often accidents also occur because workers have forgotten, after several years of unemployment, how to handle their tools and machines. This probably explains the unusual increase in 1933.

In spite of the rise in industrial sickness, the expenditure for sickness pensions decreased by about one-half (see Table 31). Disability pensions were only slightly larger notwithstanding the enormous increase in accidents. Increased expenditure on widows and orphans occurred in part because the number of

[23] *W. und S.*, XVIII (1938), 428, and XIX (1939), 422.
[24] *W. und S.*, XIX (1939), 576.

workers killed by accident had risen so rapidly. The reduction in old-age pensions is spectacular. As a result of the law of May 17, 1934, the old-age and incapacity pension systems were reorganized, but the scheme has not yet been put into operation. It was not deemed expedient to burden industry with a heavy scale of contributions. Despite growth in population, total expenditure of all kinds under all provisions for social insurance was 4 per cent less in 1938 than in 1929.

Education has also been restricted. The National Socialist government has checked the tendency to a universal academic standard of education which, despite hard times, had been growing since the first World War. Quotas were set at universities, technical high schools, and commercial institutes. Preference for access to the restricted university opportunities is given according to a scale on which membership in the National Socialist party and in the various youth organizations plays a leading part. Data concerning number of schools, classes, students, and book production, presented in Table 32, indicate the extent of reduction in educational facilities.

In summary, it may be said that the foregoing analysis emphasizes that such planning as is attempted in the Nazi economy has not redistributed wealth and income nor created a new balance among the sectors of the national economy. Thus, so far as factors connected with the propensity to consume are indigenous to crises, Nazi planning is not aimed to reduce or eliminate factors leading to another depression. Rather, the mechanism of the state is devoted "to making things go." To that purpose, the Reich Food Estate was set up, cartel supervision was tightened, a price commissioner was appointed, currency and credit control boards were introduced, and about five billion RM were spent on work creation and public works. For this purpose the Nazis abolished collective bargaining and the right to strike, established a system of government-administered prices, and decreed the numerous restrictions of the two four-year plans. In addition to the end of "making things go," the creation of a strong land, air, and sea power has been a dominating idea of public life since Hitler's rearmament declaration of March 16, 1934. National Socialism thus demanded

TABLE 32
EDUCATION IN THE THIRD REICH

	1929	1931	1934	1936	1938
I. Elementary school					
1. Schools	53,417	....	....	52,370	51,118
2. Classes	197,984	....	....	192,600	187,312
3. Students	7,708,000	....	....	7,892,000	7,621,220
4. Teachers	....	192,800	....	184,900	180,323
II. Intermediate school					
1. Schools	....	1,472	....	1,275	....
2. Classes	....	9,300	....	8,800	....
3. Students	....	229,700	....	235,200	....
4. Teachers	....	11,500	....	10,400	....
III. High school					
1. Schools	....	2,478	2,428	2,319	....
2. Classes	....	27,675	26,105*	25,896	....
3. Students	....	778,400	672,600	673,100	....
4. Teachers	....	44,900	45,300	42,700	....
IV. University (Winter semester)					
1. Schools	73	79	70	73	....
2. Students	122,400	130,100	89,100	81,400	....
3. Professors	....	....	8,271	8,188†	....
New book production	27,002	24,074	21,601	23,654	25,439

* Data for 1935, none given for 1934.
† Data for 1935, none given for 1936.
SOURCE: *Stat. Jahrb.*, 1938, p. 637, and *W. und S.*, XIX (1939), pp. 287 and 778.

sacrifices from the German people in order to achieve the "nation in arms."

The government interference has radically changed one thing: it has altered the means and methods through which the struggles between social groups are carried on and through which economic conflicts are adjusted. Opposition is no longer expressed by political elections, parliamentary debates, collective bargaining, and other democratic methods, for the state is the omnipotent arbiter over the economic classes. As the preceding discussion shows, the government has thus far settled social and economic conflicts in a manner which bestows the most important economic benefits on the larger units in industry and agriculture.

CONCLUSION

THE DEVELOPMENT of a closer dependence between the state and business was one of the profound effects of the World War and its aftermath — effects steadily at work in all leading industrial countries, though scarcely recognized until a vast cumulative change had been wrought. During the war, business and government had been compelled to work together, in new organizational forms; and the huge war-time credit expansion and general dislocation of industry left post-armistice problems which enforced further collaboration. The state and business could not be separated following the war, for the state had to care for the unemployed while business sought subsidies, the regulation of cutthroat competition, and even the "socialization" of debts. To all appearances two eras are over — the period in which business could entirely dominate government and the period in which business was largely independent.

Although this situation exists throughout western capitalism, separate heritages make capitalism appear in different guises under Fascism, Nazism, British "Tory Socialism," or the American New Deal. The essence of the Nazi ideology is the elaboration of a "mythos," made of legend and histrionics, to promote and uphold faith in the Nazi party state. And state supremacy received support from various business interests because of subsidies and special considerations given to particular groups by the Nazi state. As time went on, and Nazi power became consolidated within Germany, party members themselves appeared as serious business rivals.

Despite wide divergence in the character of their natural resources, as well as in extent of industrialization, Fascism in Italy and Nazism in Germany have striking similarities. Business in Italy also accepted the "strong man," if with misgivings. And for a time Mussolini did pleasant things for business — suppressed independent trade unions and prohibited strikes. The substantial interventions of the fascist state in the business economy took the form of financial aids and subsidies to big business. The state arranged higher protective tariffs to benefit special interests; facilitated the merger of corporations especially to save weak concerns from going to the wall; created public institutes to take over the shares of bankrupt companies until they were again in a healthy condition; forbade the erection of new factories likely to cut profits for those already in existence; relieved business from the painful and difficult operation of cutting wages; spent huge sums for military purposes, and thus provided profitable "business" for the heavy industries. Whatever the hopes expressed in rhetoric, the corporative state in Italy protected business men engaged in monopolistic enterprises and left the rest to engage in the struggle for existence under a heavy burden of taxes. Having suppressed all workers' organizations, Mussolini provided entertainment for workers after hours, " 'dopolavoro,' to occupy their minds." A large part of the money for these diversions came from indirect taxes borne mostly by the beneficiaries of *dopolavoro*. Workers were given an opportunity to organize new groups, controlled by business and Fascist party members. Although employees were provided the formal opportunity to present grievances, all disputes were decided upon by tribunals with the "proper" point of view.

The corporative state involved little more than a more complete set of trade associations familiar to other countries and in any case did not substitute public ownership for private in business enterprise. The fascist government, however, did assist large-scale business at the expense of small; and this may be a natural trend of bureaucratization of an economy. The squeezing-out of the little man meant that more weight was won by the big landowners and the army and bureaucracy.[1]

Since the seventeenth century, business in both Italy and Germany has lagged behind the business systems of other nations in its political aspects. The contemporaneous awakening of the two countries to nationalism and industrialism was late and occurred under similar circumstances. Unification in both cases was achieved by a military aristocracy — desired, however, by business liberals. The World War brought similar fates to both countries.

In Italy and in Germany the heavy industrialists undoubtedly provided encouragement and support for the anti-democratic movement. Other businesses probably followed along.[2]

Unhampered by disruptive antitrust legislation, German concerns had proceeded easily with rationalizations, mergers, consolidations, and cartelization, particularly after 1923. But the swift unfolding of the marvelous technical triumphs of German business left practically unaltered the small business men in small towns. As in Italy, there was thus a heritage of the past upon which the myths of the Nazi movement could play.

Once the Hitler government had obtained power, it exhibited features similar to those of the Italian system. Indeed, in Germany, the changes undergone by capitalism were greater, but in broad outline the two economies have remarkably similar features from an organizational point of view. The German

---

[1] William Welk, *Fascist Economic Policy* (Cambridge, Mass., 1938); Carl T. Schmidt, *The Corporate State in Action* (New York, 1939); and *The Plough and the Sword* (New York, 1938).

[2] For example, Hitler was backed by: "a seeker of a cigarette monopoly who, in fact, attained much of his desire through Hitler; a piano-maker's wife, Helen Bechstein, afterwards rewarded by an official injunction upon Germans to remember pianos; and two brothers, manufacturers of peroxide, who may or may not have been actuated by the hope of making all Germans blonde again. And in the rear hovered the vague, diminutive shapes of many small business men, representatives of a medieval economic order that had managed miraculously to survive in Germany" (Miriam Beard, *History of the Business Man*, New York, 1938, p. 747).

CONCLUSION

dictatorship aimed to prune capitalism of its "irresponsible individualism," assure it a steady return, and insure it against labor difficulties. The old employers' organizations have been set up as government-supervised associations in which the various economic groups compete for the distribution of the national income. No one is allowed to stay out and spoil the economic agreement. The functions of the old organizations have been taken over, and control has been centralized and simplified. The significance of the control machinery thus set up cannot be determined by a mere description of its structure; the important thing is how it is used and in whose interests it is operated. Numerous compulsory cartels have been formed to prevent price-cutting. The present cartel act, moreover, has not protected the individual entrepreneur if he has practiced under-cost selling.

The government, it is true, determines price and investment trends for the whole economy in accordance with the needs of the state. From the beginning, the building of an efficient war machine has been the dominating element within the economy, but within these limitations the system of private ownership of the means of production still operates.

The general principle behind price policy is that selling price should just equal the necessary costs of production, including a reasonable allowance for depreciation and a reasonable profit. A similar standard is applied in agriculture, where a comprehensive marketing organization has been set up. A "guaranteed" price is received by the farmer, while prices in distribution are determined by production costs plus distribution costs together with a standard rate of profit. This gives a considerable degree of latitude, because standards of "reasonable" profit and reasonable allowance for depreciation have never been determined. The exceptions which were granted are reflected in the rise of prices. This rise would undoubtedly have been greater had it not been for the Price-Stop Decree, but it shows that the Commissioner carried out compromise negotiations with cartels and the various groups in the Estate of Industry and Trade.

A certain amount of freedom was left to the individual business man because the officials lacked knowledge concerning his cost conditions and were reluctant to run him out of business. Many of the price decrees have stressed the necessity for proper bookkeeping and cost accountancy. The economic groups have been expressly required to foster and encourage bookkeeping in order that firms might know what their costs of production actually were, with the aim of eliminating unnecessary price rises.

The price control established under the Nazis is similar in many respects to the industrial policies of the N.R.A. The legal sanctions of the N.R.A. forced into the code authorities most of the producers in each industry and permitted the more direct attainment of the aims of the trade associations, which had striven with little success to regulate price competition. The attempt under the N.R.A. to prohibit sales at prices below the costs of production is also similar to the aims of German price policy.

Although it is true that the extent of investment control was far-reaching, its significance and incidence are frequently misunderstood. The government controlled investment by its allocation of foreign raw materials and by its control over the capital market. Firms which made profits could expand by ploughing back profits, but new issues on the capital market were restricted to firms which contributed to the rearmament and autarchic aims of the state. The effect of this was considerably to eliminate the competition of new firms, to prevent new "overinvestment" in particular industries, and to allocate investment in accordance with government objectives. In order to encourage private industries under the Second Four-Year Plan, the state in many cases guaranteed to the firms concerned a "reasonable" profit if they would undertake the desired production, and if firms lowered costs they were not forced to lower prices.

The preceding chapters have indicated the enormous complexity of state regulation and interference. It needs to be emphasized, however, that the National Socialist government is opposed to state ownership if the interests of a strong state can be preserved without it. As has been shown, many undertakings in fact were "denationalized" and "de-municipalized." The controls which have been set up — for example, in the banking system and the money market — have attempted to leave as much as possible of the old system of private ownership of the means of production. The general effect of these policies would seem to be to strengthen the position of heavy industry. Large business enterprises which may have found it difficult in the past to prevent price discrimination and competition without ownership can now use the arm of the state to enforce price policies. This, of course, undoubtedly involves the expense of lobbying, intrigue, and pressures upon party members.

Nazi Germany has quite generally been seen as the world's foremost nesting ground of economic paradoxes. Some observers have believed that its economic system was devised as the last refuge of monopoly capitalism, though industrialists took orders from generals. Nazi Germany has been labeled "Brown Bolshevism." Yet the profit system is one of its integral parts; the inequality of distribution of wealth and income has increased, and wages have been kept at depression levels, while profits have considerably recovered. Observers have been predicting "a runaway inflation" for the last six years, but events have belied their prognostications. These paradoxes, however, are resolved when one remembers that Germany has been arming herself almost since the beginning of the Nazi regime. It is the concept of the totalitarian war which explains the nature of the German economic system.

Totalitarian preparedness and the consolidation of Nazi

power at first necessitated a reëmployment program. The considerable reserve of unemployed productive power available in Germany at the beginning of the new regime as a result of the preceding extreme deflationary policy was used for militarization. Employment was increased by public investment, and inflation was prevented by restricting the increase of consumption and private investment consequent to state expansion. Totalitarian preparedness also required a program of self-sufficiency in food and materials so that Germany might be able to withstand a wartime blockade.

In order to carry on the gigantic program, the government needed the support of private capital, particularly in heavy industry. In order to make Germany self-sufficient, the government continued protection of the large agricultural estates, which were of great importance for the grain and meat supply of the nation. The continuation of the military plans, however, meant that the state tightened control over production and distribution.

The Nazi government was successful in evolving a system — initially empirical — based on three main controls: of costs, investment, and international trade. The considerations presented here seem to demonstrate that this system is not subject to wide fluctuations in employment so long as the available powers of control are ruthlessly and skilfully used. Rearmament had the most stimulating effect on employment, output, and incomes; at the same time, the growth of investment created the incomes by means of which the works, of whatever kind, were carried out. Without this increase in investment a great share of the incomes themselves would never have existed. As soon as a state of full employment was reached, the situation was fundamentally changed. Apart from technical progress, innovations, and conquests, further investment could not increase incomes. From full employment on, it became true that more of A would have to be produced at the expense of less of B.

CONCLUSION

This in turn involved further state control. The position of the German business men, however, was considerably eased by the military conquests of 1939 and 1940.

Difficulties in financing do not seriously threaten the system. Conservative methods of public financing have been greatly emphasized, so that almost a half of armament expenditures come out of taxation. So long as loan expenditure is confined to the limits of enforced and voluntary saving it cannot lead to cumulative inflation, however financed. On the other side, the government is not obstructed by the resistance of investors. Export of capital is impossible, and a flight into commodities is impracticable. Although the investor can depress the rate of interest on non-government securities below the rate of interest on government securities, the strict control of new issues and of construction prevents private investment from competing with public investment. The government can offset hoarding on the part of private investors by financing state requirements to a greater extent through rediscounting at the Reichsbank. Institutional savings are under direct control and cannot hold out for higher interest rates.

When full employment is reached, the extent of what may be termed barren loan expenditure also appears at first sight to be reaching a maximum, because the service of the debt burden rises, while revenue does not expand. The government can deal with this problem by lowering the rate of interest. Moreover, a high debt burden is detrimental only if the additional taxation caused by it has a depressing effect on expectations of future yields on new capital investments. If this depressive effect does not exist, the scale of activity remains unchanged; and in Germany, where the government subsidizes and controls the capital market, a considerable reduction in net yield would still not result in investment sabotage. Moreover, the debt is financed by a regressive tax system, which increases existing inequalities in income, thus increasing the propensity to save and thereby assisting the government's task of restricting consumption.

The short-run achievement of the Nazis is evident. The use of moral and material resources which in other countries are usually not mobilized before the start of hostilities has yielded impressive results as measured in an economy on a war footing. A German victory, moreover, may enable the standard of living to rise, since the exploitation of foreign workers and capitalists would enable German profits to rise at the same time that an increase in German wage rates took place.

The highly imperialistic character of the Nazi system arrives at completeness in its economic sphere. Individual Nazi party members are enabled to use the armed forces of the state for the aggrandizement not only of the one-party state but of individual members of the party as well. The increase in wealth and income of Nazi party members is somewhat analogous to the loot and booty of feudal robber barons. The analogy breaks down because the feudal robbers did not have the weapon of the national state; Nazis have not only the state but likewise all the equipment and industrial organization developed since the Industrial Revolution.

The new industrialists of Germany are Nazi party members, and their competitors — other business interests at home and abroad — are at an extreme disadvantage. These non-Nazi business interests are similar in many respects to small businesses which were crushed or swallowed up by larger interests. What was once considered mammoth in business has been dwarfed by the possibilities open to individuals who may use not only armed forces but also the whole framework of government as represented by legislature, judiciary, and executive. Whereas earlier imperialist business interests were impeded and retarded by other elements in the society, the Nazis have set up a system whereby imperialist business interests may pursue their objective practically unimpeded by pressure from other groups. The retention, in part, of a system of private ownership enables party members to build for themselves industrial empires; at the same time, restrictions upon private property, such as government determination of investment and regulation of prices and wages, make it possible for the economy to function as a war weapon without interference from rival industrialists or from consumers and workers. The owning class still exercises one function — the receiving and accumulating of profits. But industrial concentration and the increasingly pervasive influences of the bureaucracy have given a new aspect to the "capitalist" order. A new parasitic group — Nazi party members belonging for the most part to the old middle class — penetrates more and more into the realm of property. Tyranny in the age of machines presages the continuation of an economic system which is basically decadent but efficient and aggressive in war.

BIBLIOGRAPHY

BIBLIOGRAPHY

This bibliography includes only books and articles cited. The literature on Germany is so extensive that it seems necessary to limit the list to materials actually used.

BOOKS, PAMPHLETS, AND ARTICLES

Balogh, Thomas, "The National Economy of Germany," *Economic Journal*, September 1938.

Barberino, Otto, "Veraenderte Struktur des oeffentlichen Haushalts," *Der Wirtschaftsring*, July 22, 1938.

Beard, Miriam, *History of the Business Man* (New York, 1938).

Billich, Carl, "Vier Jahre nationalsozialistische Kartellpolitik," *Der deutsche Volkswirt*, 1936, p. 2527.

Bowley, Arthur, *The Changes in the Distribution of the National Income, 1890–1913* (Oxford, 1920).

Brady, Robert, *The Spirit and Structure of German Fascism* (London, 1937).

Brandt, Karl, "The German Back-to-the-Land Movement," *Journal of Land and Public Utility Economics*, May 1935.

——, "Junkers to the Fore Again," *Foreign Affairs*, October 1935.

Bresciani-Turroni, Costantino, "Annual Survey of Statistical Data: Pareto's Law and the Index of Inequality of Income," *Econometrica*, April 1939.

Brinkmann, Rudolf, *Staat und Wirtschaft* (Stuttgart and Berlin, 1938).

Clark, Colin, *Conditions of Economic Progress* (London, 1940).

Dessauer, Marie, "The German Back Act of 1934," *Review of Economic Studies*, June 1935.

Ellis, Howard, "Exchange Control in Germany, *Quarterly Journal of Economics*, supplement, vol. LIV (1940).

Ermarth, Fritz, *The New Germany* (Washington, D. C., 1936).

Feder, Gottfried, *Das Manifest zur Brechung der Zinsknechtschaft* (Munich, 1932).

Great Britain, Department of Overseas Trade, *Economic Conditions in Germany to March 1936* (London, 1936).

Guerin, David, *Fascism and Big Business* (New York, 1939).

Guillebaud, C. W., *The Economic Recovery of Germany* (London, 1939).

Hamburger, L., "How Nazi Germany Has Mobilized and Controlled Labor" (pamphlet, Brookings Institution, 1940).

Higgins, Benjamin, "Germany's Bid for Agricultural Self Sufficiency," *Journal of Farm Economics*, May 1939.

Hugenberg, Alfred, *Streiflichter* (Berlin, 1927).

Lachmann, Kurt, "The Hermann Goering Works," *Social Research*, February 1941.

Laufenburger, Henri, and Pierre Pflimlin, *La Nouvelle Structure économique du Reich* (Paris, 1938).

League of Nations, *Public Finance, 1928–35: XII (Germany)* (Geneva, 1936–39).

——, *Money and Banking*, II, 1938 (Geneva, 1939).

Lehmann, Fritz, and Hans Staudinger, "Germany's Economic Mobilization for War," *The Conference Board Economic Record*, June 24, 1940.

Linhart, Elisabeth, "Wettbewerbstheorien-Wettbewerbspolitik," *Vierteljahrshefte zur Wirtschaftsforschung*, Heft 1, 1939/40.

Muellensiefen, Heinz, *Gruppenaufgaben bei der Wirtschaftlichkeitsfoerderung, Marktordnung und Kartellaufsicht* (Stuttgart, 1937).

——, *Kartellrecht einschliesslich neuer Kartellaufsicht,- Preisbildung,- Schiedsgerichtsbarkeit und Steuerrecht* (Berlin, 1938).

——, *Vor der Kartelpolitik zur Marktordnung und Preisueberwachung* (Berlin, 1935).

Poole, Kenyon, *Financial Policies in Germany, 1932–1939* (Cambridge, Mass., 1939).

Raphael, Gaston, *Le Roi de la Ruhr, Hugo Stinnes* (Berlin, 1925).

Reimann, Guenter, *The Vampire Economy* (New York, 1939).

Salewski, W., *Das auslaendische Kapital in der deutschen Wirtschaft* (Berlin, 1930).

Salvemini, Gaetano, *Under the Axe of Fascism* (New York, 1936).

Schacht, Hjalmar, *The End of Reparations* (New York, 1931).
Schattendorn, Karl, "Konzern Hermann Goering Werke," *Der Wirtschaftsring*, December 23, 1938.
Schiller, Karl, *Arbeitsbeschaffung und Finanzordnung in Deutschland* (Berlin, 1936).
Schmidt, Carl T., *The Corporate State in Action* (New York, 1939).
——, *German Business Cycles, 1924–33* (New York, 1934).
——, *The Plough and the Sword* (New York, 1938).
Schmoller, Gustav, "Die Einkommensverteilung in alter und neuer Zeit," *Jahrbuch fuer Gesetzgebung, Verwaltung und Volkswirtschaft* (1895).
Singer, H. W., "The German War Economy in the Light of German Economic Periodicals," *Economic Journal*, December 1940.
Stamp, Sir Josiah, *Wealth and Taxable Capacity* (London, 1930).
Steinberger, N., *Die Agrarpolitik des Nationalsozialismus* (Moscow: Internationalen Agrarinstitut, 1935).
Stolper, Gustav, *German Economy, 1870–1940* (New York, 1940).
Stresemann, Gustav, *Vermaechtnis* (Berlin, 1932).
Sweezy, Maxine, "German Corporate Profits: 1926–38," *Quarterly Journal of Economics*, May 1940.
——, "Distribution of Wealth and Income under the Nazis," *Review of Economic Statistics*, November, 1939.
Untersuchung des Bankwesens (Berlin, 1933–34).
Vollweiller, Helmut, "The Mobilization of Labour Reserves in Germany," *International Labor Review*, November 1938.
Welk, William, *Fascist Economic Policy* (Cambridge, Mass., 1938).
Wilde, J. C., de, "Germany's Wartime Economy," *Foreign Policy Reports*, June 15, 1940.

PERIODICALS, NEWSPAPERS, YEARBOOKS

Arbeitsrecht
The Banker Magazine
Deutsche Bank, *Wirtschaftliche Mitteilungen*
Deutsche Bergwerkszeitung
Deutsche Freiheit
Der deutsche Volkswirt

Deutscher Reichs- und Preussischer Staatsanzeiger
Deutsche Wirtschaftszeitung
Economist (London)
Einzelschriften zur Statistik des Deutschen Reichs
Frankfurter Zeitung
Institut fuer Konjunkturforschung, *Konjunkturstatistisches Handbuch*
——, *Statistik des In- und Ausland*
——, *Vierteljahrshefte zur Konjunkturforschung*
——, *Weekly Report*
——, *Wochenbericht*
International Labor Review
Journal du Commerce
League of Nations, *Monetary Review*
——, *Statistical Yearbook*
London Times
Monthly Labor Review
News in Brief
New York Times
Reichsarbeitsblatt
Reichsgesetzblatt
Reichs-Kredit-Gesellschaft, *Germany's Economic Situation*
Soziale Praxis
Statistik des Deutschen Reichs
Statistisches Jahrbuch fuer das Deutsche Reich
Le Temps
Der Vierjahresplan
Vierteljahrshefte zur Statistik des Deutschen Reichs
Voelkischer Beobachter
Der Wirtschaftsring
Wirtschaft und Statistik
Zeitschrift fuer oeffentliche Wirtschaft

INDEX

INDEX

Symbols attached to page numbers have the following meaning: c, chart; n, footnote; t, table.

Advertising, 105
Agriculture, "battle of production, 22-23; chaotic market, 7; decline of raw material imports, 123t; Hereditary Farm Act, 180–183; land ownership, 179–183, 182t; income consumed on farm, 200; Reich Food Estate, 183–189; regulation of prices, 184–187; rural and suburban settlements, 189–193
Agricultural Adjustment Administration, 188
Allgemeine Deutsche Kreditanstalt, 31
Alpine Montangesellschaft, 87
Armaments, effect of expansion, 23; expenditure on, 160t, 204; goal of Nazi planning, 24. *See also* War measures
Asbestos industry, dividends in, 81t; profits in, 72t, 76
Asia, German imports to, 122t
Aski marks, 119–120
Association of Chambers of Commerce and Industry, 40
Association of German Industry, 112
Austria, 86–87
Austrian Creditanstalt, 83, 86
Austrian Industrie-Kredit, A.G., 84
Automobiles, 51, 223. *See also* Conveyances; Transportation

Balkans, 115, 121, 122t
Balogh, T., 20n
Banking, size of firms in, 81t; bankruptcies in, 82t; government ownership before Nazis, 29–30; profits in, 73t, 79; reforms, 134–141
Bankruptcy, 65, 82t
Barberino, O., 157n
Barter deals, 118–119. *See also* Clearing agreements

Bau, A.G. Negrelli, 86
Beard, Miriam, 232n
Belgium, 114n
Benzolvertrieb der Reichswerke Hermann Goering, A.G., 87
Bilateral trade agreements, *see* Clearing agreements
Billich, Carl, 103n
Bond price, 127, 129
Borsig, Herr, 35
Bosch, Karl, 35
Bowley, A., 219n
Brady, Robert, 41n, 178n
Branded goods, 97, 100. *See also* Trade marks
Brandt, Karl, 180n
Bresciani-Turroni, C., 220n
Brinkmann, R., 44, 45–46, 209n
Brno arms factory, 87–88
Bruening, 9, 10, 126, 180
Building, construction in Germany, 223t; materials, 73t, 76, 78, 100; trade, 73t, 76, 78, 81t; restriction during war, 107. *See also* Housing

Capacity, 51–52
Capitalism, 27, 230–240
Capitalists, 28, 239
Capital flight, 110–112
Cartels, 90–98; Compulsory Act, 92–93; price-control function, 94–98
Central Europe, 100n, 109n
Chain stores, 46–47
Chamber for Culture, 53, 56
Chambers of Commerce, 40, 41n, 41–42, 52
Chemical industry, decline in mineral oil imports, 123t; dividends, 81; mergers, 83; plant expansion limited, 92; profits, 73t, 78; reduction of

INDEX

branded goods prices, 100; size of corporations, 81t
Clark, Colin, 203n
Clearing agreements, 115–116, 117, 118–119, 121
Clothing industry, 72t, 222–223
Combat League of the Middle Class, 36
Commerce, profits in industry, 72t
Commerz- und Privatbank, 31–32
Competition, attitude of Nazis toward, 85; limited in shipping industry, 61; regulated by decree of December *1934*, 97; restraint of, 45–47; restricted by Cartel Act, 93; views of German economists on, 103
Consumers' goods industry, bankruptcies, 82; curtailment under New Plan, 20; size of corporation, 81t. *See also* Light industries
Consumption, limited recovery of, 220–224; rationing, 106; regulated by Council for Defense, 52
Conversion, 126–134
Conveyances, 72t, 76
Coöperatives, 47
Copsa Mica, 88
Corporation of German Industry, 36, 37
Corporations, bankruptcies, 82; centralization of control, 66–70; Civil Code of *1884*, 66; control for state aims, 69–70; dividends, 81t; law of July 5, *1934*, 64; decree of October 9, *1934*, 64; law of *1937*, 63–70; profits, 70–80; size of, 81t, 81–82; supervisory board of Rheinmetall-Borsig, 35; tax, 153–154, 198n
Corporative structure, 36–38
Cost of living, 10, 201, 201n
Credit regulations, 136–139
Credit supervisory board, 135–136
Czechoslovakia, 87–88
Czechoslovakia Živnostenská Banka, 84

Darre, Walter, 179n, 180
Dawes loan, 113, 114
Deficit financing, 145–150. *See also* Government expenditure; Government financing; Government revenue
Deflation, 8–11
Depreciation accounts, 53, 80
Depression, 7–10
Dessauer, Marie, 135n, 141n
Deutsche Bank, 31–32, 35, 83, 84
Deutsche Bau- und Bodenbank, 30
Deutsche Rentenbank, 30
Deutscher Schiff und Maschinenbau, 32
Dietrich, Finance Minister, 30
Distribution, of income by size class, 28, 211–216, 212t; index of inequality, 212t; of salaries, 215–216, 216t; of wages, 215–216, 216t; of wealth, 28, 216–220, 217c. *See also* Profits; National income
Dresdner Bank, 31, 32, 35, 83, 84, 88

Education, 227, 228t
Efficiency, 85, 104, 105
Electrical industry, 51, 72t, 83, 100
Ellis, H., 108n, 109n, 123n, 124n
Employers' organizations, 35–39
Employment program, 7–25; pre-Hitler measures, 7–12; work creation and supposed remedies for unemployment, 12–18; rearmament, 18–21; Second Four-Year Plan, 21–24
Ermarth, F., 188n
Estate of German Handicrafts, 47–48
Estate of Industry and Trade, 29–44, 47, 53, 54, 55, 56; encourages efficiency 85; duties of, 102–104; export levies determined by, 121; relation to cartels, 93–94, 101–102
Europe, western, 122t
Excise, 156
Export levy, 121

Fanto, A.G., 87
Farbenindustrie, I.G., 35
Fascism, compared to Nazism, 230–232
Feder, Gottfried, 32n
Feinstahlwerke Traisen, A.G., 86
Ferrous metals, decline of imports, 123t; dividends in iron and steel in-

INDEX

dustry, 81; mergers, 83; profits in iron and steel industry, 72t, 76; reprivatization of Steel Trust, 28, 30–31. *See also* Heavy industry
Financial holding companies, 72t, 81t
Fine mechanics industry, 72t, 76, 77, 92, 100
Finishing industry, 64
Fishing industry, 72t, 76
Flight from the land, 192
Foodstuffs, profits rates in industry, 72t
Foreign exchange control, administrative provision, 109; boards, 99, 117–118; defined, 109; goal of, 108–109; totalitarian institution, 117–124
Foreign exchange rates, 112–117
Foreign investment, in Germany, 7n
France, 114n
Full employment, 236, 237
Funk, W., 49, 50, 53, 131

Gainfully occupied, 207t
Gelsenkirchen Mining Co., 30–31
Gemeinnutz vor Eigennutz, 43–44
German African Lines, 60–61
German Federation of Industry, 36, 38, 40
German Levant Line, 60–61
German Steel Trust, 30–31, 87
Gewerkschaft Sachsen, 88
Goering, Albert, 88
Goering, Hermann, 21, 34, 48, 49, 85; appoints leaders of defense industries, 51; ascendancy to power, 98; commander of all economic activities, 53; opposes nationalization of industry, 50; power to control prices, 98
Goering Works, Hermann, 34, 35, 83, 85–89, 95
Gold Discount Bank, 30, 31, 113, 114, 121, 147
Goltz, von, 37
Government debt, 145–151, 157–160, 158t
Government expenditure, 11, 145–150, 158t, 160t

Government financing, effect on profits of banks, 79; conversion, 126–134; funding operations, 147–150; reduction rate of interest, 126–134; summarized, 235–237; war measures, 149–150
Government ownership, history of, 28–30; restricted in electric power industry, 32; trends in, 27–35
Government revenue, 150–157, 159t
Great Britain, 114n, 122t
Guillebaud, C. W., 34n, 100, 130n, 133n, 188n
Guerin, D., 181n, 182n

Hamburg-America Line, 60–61
Hamburg-South America Steamship Company, 60–61
Hamburger, L., 172n
Handicrafts, 47–48, 173
Hanneken, von, 49
Heavy industry, bargaining power, 84; bankruptcies in, 82; depreciation allowances in 1938, 80; profits in, 72t, 76, 77, 79; new firms, 82; newspaper of, 34; size of corporations in, 81t
Hereditary Farm Act, 180–183
Higgins, B., 185n, 188n
Hitler, alluded to, 7, 12, 17, 27, 36, 51, 58, 63, 96, 113, 126, 229; capitalism upheld by, 26; mass leader, 27; opposes nationalization of banks, 31–32; promise to industrialists, 26; publishing industry owned by, 85; secret of successes, 4
Holding companies, profits, 73t, 79
Holland, 114, 114n
Housing, 10, 12, 190–191, 223–224, 223t
Hugenberg, A., 27n, 180

Imperialism, 238
Imports, decline in, 123t; distribution among countries, 122t; totalitarian control of, 117–124
Income tax, 152–153, 198n
Inflation, 3, 9, 39, 236–237. *See also* Government financing

Inheritance tax, 154
Inequality, of income, index, 212t; of wealth, index, 218t. *See also* Distribution; Profits; National income
Insurance companies, dividends, 81; profits, 73t, 79; size of, 81t
Interest rate, reduction of long-term, 126–134
Investment, government control of, 234–235; multiplier, 17–18. *See also* Employment program
Investment companies, profits, 73t, 79
Italy, 114n

Jawitz, von, 49
Jews, 35, 47, 88, 172, 179–180, 181
Journeymen, 47
Junkers, 192

Kessler, Minister of Economy, 37
Krupp, 36, 199

Labor, apprenticeship, 174–176; conscription, 52, 176–177; convicts, 172–173; Courts of Social Honor, 165–166; definition of wage income, 215; distribution of salary income, 216t; distribution of wage income, 216t; feudalism, 169; forty-hour week, 13; income of, 208–211; mobilization of, 171–174; multiple earnings, 15; Mutual Trust Councils, 163–164; neutralization of, 191; overtime wage rates, 106; regulation of working conditions, 165–166; salary income, 208t; spreading of work, 12–13; "Strength through Joy" activities, 211t; tax burden, 209, 210t; trustees, 164–165; wage income, 208t; workbook, 166–167. *See also* Employment program
Labor Front, 36, 38, 53, 56, 58, 161–163
Labor Service, 167–168
Lachmann, K., 86n, 87, 88n
Laenderbank, 84
Lange, Herr, 48
Laufenburger, H., 42n
Leather industry, profits, 72t

Lehmann, F., 121n
Ley, R., 36–37, 58, 85
Light industry, 72t, 75, 77
Linoleum industry, profits, 73t
Linhart, Elisabeth, 103n
Liquidity of banking system, 139–141
Living costs, *see* Cost of living
Loan Fund Act, 67, 130–134
Loeb, Major-General, 49
Luxuries, bankruptcies, 82; expanded to offset necessities, 223–224; profits, 73t, 75, 76, 77; rationing of, 106; size of corporations making, 81t

Machine industry, 51, 52, 72t, 76
Maschinen- und Waggonbau Fabrik, A.G., 86
Mergers, 64, 83. *See also* Goering Works, Hermann
Mining, coal, 94–95; dividends, 81; government commission of *1919*, 29; government regulation, 29; profits, 72t, 78
Minister of Economic Affairs, 34, 40, 41, 49, 53, 70, 91–92, 94
Mixed enterprises, 33, 33n, 34–35
Mobilization of labor, 168–177
Monopoly, decline in bankruptcies protects, 85; increased by government, 10, 46; protection under foreign exchange control, 112–113. *See also* Corporations; Cartels; Price control; Price policy
Motor vehicles, 57, 106. *See also* Automobiles; Transportation
Muellensiefen, 91, 92n, 93n, 94n
Municipalities, 32–33, 40, 128
Musical instrument industry, 73t, 76, 77

National Economic Chamber, 40, 41n, 48, 56
National income, causes of increases, 205–207; criticism of Nazi statistics of, 197–200; defined, 197–198; deflated by price indexes, 200–205; size of, 196–207; structure of, 207–211

INDEX

National Service Law, 52
National Transport Council, 55–56
Nazi party, "anti-capitalism," 26; attitude toward competition, 85; consolidation of power, 28; distribution of spoils, 14, 57, 195; leader of economic section in, 35; left wing, 35, 49, 127; new parasitic group, 239; rise of, 26–27, 161
New plants, 52, 54, 82
New Plan, 19, 117
Nonferrous metals industry, 97, 100, 123t
Norddeutsche Kreditbank, 31
North German Lloyd, 60–61

Oil mergers, 83
Old age, employment preference, 14; insurance pensions, 226t
Optics industry, 72t, 76, 77
Osthilfe, 192

Papen, von, 10, 27
Paper industry, 72t, 92
Pareto method, 212–216
Pauker Werke, A.G., 86
Peasant holdings, 182t
Pensions, social insurance, 226–227, 226t
Pflimlin, Pierre, 42n
Phoenix Mining Co., 30
Planning, summary evaluation of Nazi, 193–194, 227–229
Pleiger, Paul, 95
Poole, Kenyon, 11n
Preussengrube, A.G., 88
Price control, commissioner for the supervision of, 10; emergency decree of *1931*, 96; decree of July 15, *1933*, 96; decrees of August and December *1934*, 97; decree of March *1935*, 97; decree of November *1936*, 98; decree of July *1937*, 98; decree of September *1939*, 99–100; function of cartel, 94–98; German economists' attitude toward, 103; loopholes in, 101; of nonferrous metals, 97; reduction of costs by, 100; textiles under, 97, 99; war measures, 103–107
Price index, 201–204, 203n, 204n
Price policy, 96–107, 233–234
Producers' goods, size of corporations, 82. *See also* Heavy industry
Production index, 203, 203n
Profits, 10, 53, 70–80, 99, 208t
Propensity to consume, 20, 28, 220, 227
Property, defined, 218; income from, 208–209, 208t; tax on, 154, 218, 218n
Proxy voting, 69
Public expenditure, *see* Government expenditure
Public utilities, 10, 32n, 32–33
Public works, direct effect of pre-Hitler measures, 11; German case no indictment of, 19; pre-Hitler measures, 7–12; Schleicher's program, 11; under Hitler, 16–18; under von Papen, 10. *See also* Employment program
Publicity of corporate activities, 68–69

Radios, 223
Railroads, 17t, 57t, 107. *See also* Transportation
Raphael, G., 27n
Rasche, Karl, 88
Rationalization, 51, 54
Rationing, 106
Rearmament, 18–21. *See also* War measures
Reich Coal Commissioner, 94
Reich Economic Council, 39
Reich Economic Court, 70
Reich Food Estate, 40, 53, 56, 98, 183–189
Reichsbank, alluded to, 30, 53, 59, 115, 116, 117; deficit financing, 146–147; funding operations, 148; largest contributor to loan stock fund, 132; open–market operations, 129; reform of banking system, 134–141
Reimann, G., 45
Roads, 17t, 62
Rumania, 88
Russia, 122t

INDEX

Salary income, 208t
Salewski, W., 7n
Saxony-Coburg-Gotha, Duke of, 35
Scandinavia, 122t
Schacht, 3, 7n, 19, 31, 38, 49, 107, 117, 134
Schattendorn, K., 86n
Scheer, Herr, 49
Schiller, Karl, 17n
Schleicher, 11, 127
Schmidt, C. T., 10n, 232n
Schmidt, Guido, 88
Schmitt, Minister of Economy, 36–37
Schmoller, G., 220n
Scrip system, 113–114
Second Four-Year Plan, 21–24, 48
Self-financing, 89, 131
Share prices, 126–127
Shipping, 59–62
Sickness, 224–226, 225t
Singer, H. W., 104n, 105n
Skoda Works, 87–88
Small business, 46, 65. See also Handicrafts
Social Democrats, 179–180
Social insurance, 226–227, 226t
Social welfare, 224–229
South America, 115, 119, 120, 122t
Stamp, Sir Josiah, 219n
Standstill agreement, 111
Statistics, general criticism of German, 5–6
Staudinger, H., 121n
Steel, see Ferrous metals
Steinberger, N., 193n
Steirische Gusstahlwerke, A.G., 86
Steyr-Daimler-Puch, A.G., 86
Stinnes, H., 27n
Stock exchange, function of, 144–145; regulation of, 141–145; speculation, 65
Stockholders, 67–68, 69
Stolper, G., 29n, 30
"Strength through Joy," see Labor
Stresemann, G., 27n
Subsidies, 14–15, 59, 104, 140, 191, 192
Sudetenlaendische Bergbau, A.G., 87, 88

Sweden, 114
Sweezy, M., 71n, 211n, 215n
Switzerland, 114, 114n

Tax certificates, 10–11
Taxes, alleviation of, 16; corporation, 153–154, 198n; evasion of, 200; excise, 156; income, 152–153; increased under Bruening, 9; inheritance, 154; of a working-class family, 210t; property, 154; rebates, 10; stock exchange sales, 144t; turnover, 155; wage, 154; war, 104
Textiles, new plants limited in, 92; price control of, 97, 99; profits, 72t; rationing, 106
Thomas, General, 25
Thyssen, F., 30, 88
Trade agreements, see Clearing agreements
Trade marks, 105. See also Branded goods
Transfer moratorium, 11, 113, 114
Transportation, government ownership in, 55; organization of, 55, 56; profits of private business in, 73t, 76, 78; railroad rates, 10, 56; railroad mileage, 55t; represented in war council, 53; rolling stock, 57t; shipping, 59–62; traffic carried, 57t; Volksauto, 57–59; war measures, 62
Trendelenburg, State Secretary, 35

Undistributed profits, 208t
United States, 115, 119, 122t

Van der Zypen Steel Co., 30
Veblen, T., 4, 5
Veitscher Magnesitwerke, A.G., 87
Vereinigte Industrie-Unternehmungen, A.G., 29, 84
Vereinigte Stahlwerke, A.G., 31
Volksauto, 58–59, 78
Vollweiller, H., 174n

Wage tax, 154–155, 198n
Wagener, Dr., 35–36
Wagner, price commissioner, 98, 100

INDEX

Walter, Paul, 94
War measures, army control of "w" plants, 104; armament expenditure, 159–160, 160t; coördination for self-sufficiency and war, 48–54; corporation taxes, 153–154; council for national defense, 52; dissolution of cooperatives, 47; effect on housing, 224; excise taxes, 156; expansion of Goering Works, 86–89; financial policies, 149; Germanization of foreign concerns, 83–84; government finances, 149–150; income tax, 152–153; limitation of building construction, 107; mobilization of labor, 168–177; price control, 103–107; rationing, 106–107; rearmament, 18–21; regulation of advertising, 105; regulation of trade marks, 104–105; roads, 62; strengthening of coal cartel, 94–95; Second Four-Year Plan, 21–24; transportation, 62, 106–107; wage-stop decree, 105–106
Water, gas, and electrical industry, 73t, 78, 81t
Wehrwirtschaftsfuehrer-Korp, 51
Weimar Republic, 29
Welk, W., 237n
Wholesale price index, 202, 202n
Wood industry, 73t, 76, 77
Work creation, 12–18

Young loan, 113–114